MICRODROP
GENERATION

Nano- and Microscience, Engineering, Technology, and Medicine Series

Series Editor
Sergey Edward Lyshevski

Titles in the Series

MICRODROP GENERATION

Eric R. Lee

Stanford Linear Accelerator Center
Stanford University

CRC PRESS

Boca Raton London New York Washington, D.C.

Library of Congress Cataloging-in-Publication Data

Lee, Eric R.
  Microdrop generation / Eric R. Lee.
      p.   cm. — (Nano- and microscience, engineering, technology, and medicine series)
    Includes bibliographical references and index.
    ISBN 0-8493-1559-X
    1. Atomizers. 2. Spraying. 3. Electrostatic atomization. I. Title. II. Series.

    TP159.A85 L44 2002
    660′.29—dc21   2002031317

Visit the CRC Press Web site at www.crcpress.com

Foreword

In the past two decades there has been a tremendous increase in the technological and research use of microdrops, liquid drops with diameters ranging from micrometers to several hundred micrometers. With respect to technology, one need only look at the worldwide use of inkjet printers, or for a more recent example, the growing use of microdrops for the preparation of biological microarrays. Microdrops are being used increasingly in many areas of research: microfluidics, combinatorial chemistry, biological assays, combustion science, aerosol science, and much more.

Along with this growth in microdrop research and technology, there has been a parallel growth in the microdrop technical literature: articles, patents, conference proceedings, and specialized reviews. For either the newcomer to this field or the seasoned practitioner, it is an exhausting task to become knowledgeable about the contents of this literature or to stay knowledgeable. Much of the literature is repetitive, often crucial practical and theoretical details are omitted. And in recent years, as microdrop technology has become more profitable, there has been a steady increase in the amount of information that is kept proprietary — a natural effect of the industrial success of microdrop technology, but a difficulty for the newcomer or practitioner who does not want to reinvent the wheel.

For a decade, the field of microdrops has needed a book that presents a comprehensive introduction and guide to microdrop science and technology, a book that references the useful literature, and most important, a book written by a seasoned practitioner and researcher in the field. Finally we have the book: *Microdrop Generation* by Eric R. Lee.

Lee, an engineer, researcher, and inventor, has written a book that will be valuable to everyone engaged in using microdrops in areas ranging from research to prototyping, from device design to industrial processing. This book will also be valuable for those outside the world of microdrops because it is more than a combined textbook, laboratory manual, and literature guide. It will stimulate those outside the microdrop world to think about using microdrops in their research and engineering work.

Martin L. Perl
1995 Nobel Laureate in Physics
Stanford University

Acknowledgments

The techniques, inventions, and theories of microdrop physics presented in this book are the product of nearly 10 years of experimental work, consultations with industry experts, and literature searches by many individuals from the Microdrop Particle Search Group at Stanford Linear Accelerator Center (SLAC). The material presented in this book represents only the basics of microdrop engineering and is far from the final word on the subject. We are in awe of the microdrop technology that exists in the inkjet printing industry and regret that industry inkjet engineers are not freer to publish openly the details of their practical know-how.

Our motivation for learning the art and science of microdrop generation was to implement a search for exotic, stable, subatomic particles using microdrops to introduce the test materials into our detectors. In 1992 at SLAC, Professor Martin Perl, with physicists Klaus Lackner, Gordon Shaw, and Charles Hendricks, initiated the microdrop-based search for exotic fundamental subatomic particles. Martin Perl led the research effort up to the present time, co-designed our first microdrop generators, and has compiled and synthesized or derived from first principles the majority of the theoretical insights we have of effects of fluid rheology on drop formation, the theory of drop charging, droplet evaporation and drop kinetics.

The first drop ejection hardware we worked with was a high-pressure, continuous jet microdrop generator modified to be also capable of operation in drop-on-demand mode. It was designed by Charles Hendricks, Martin Perl, and SLAC mechanical engineer Gerard Putallaz. It was brought to useful operation by the efforts of Martin Perl, and then physics graduate students Brendan Casey, George Fleming, and Nancy Mar, all of whom now have their doctorates in physics. Brendan Casey and George Fleming were from San Francisco State University (SFSU), where Professor Roger Bland's research group designed a microdrop ejector for an automated Millikan experiment. We ended up adopting for our experiments modified versions of the tubular fluid reservoir microdrop ejector design that Professor Bland and his students utilized.

The experimental studies at SLAC of the fluid properties needed for making ejectable stable suspensions were primarily the work of Professor Dinesh Loomba of the University of New Mexico, then a post-doctoral researcher at SLAC when he conducted the ejection fluid research, which resulted in a successful method of making jettable meteorite suspensions, and the work done at SLAC on the principles of colloidally suspended solids by then physics graduate student Valerie Halyo.

Our understanding of the behavior of microdrop generators and the physics of microdrops would not have been possible without the computer based digital imaging systems that were constructed and programmed by a series of physics graduate students. Our first system, which was required to perform real time image analysis using a 66 MHz, 486-based PC, was successfully programmed for this task by then graduate student George Fleming. Subsequently, Irwin Lee designed and programmed a Networked Linux, cluster-based, machine vision system to execute the more sophisticated image analysis required for our later experiments. As graduate students, Irwin Lee and Valerie Halyo extended the capabilities of our imaging systems to track multiple drops independently in real time in increasingly complex multidrop image fields, as well as to extract the drop diameter from images of diffraction blurred microdrops.

I was taught the science and art of fabricating micromachined structures during formal course work with Stanford professors Gregory Kovacs and B. (Pierre) T. Khuri-Yakub. The hands-on details of what it took to implement the principles of microfabrication on real world machines and wafers were generously and patiently given to me by the technical staff of the Stanford Nanofabrication Facility, where the micromachining of our ejection nozzles was performed.

When we started constructing microdrop generators, we were fascinated by the technology but ignorant of the diverse applications of microdrops. Our introduction into the uses of microdrops for manufacturing and basic science started with a series of personal introductions to other groups utilizing fluid microdrops by Mary Tang, biotechnology liaison at the Stanford Nanofabrication Facility, and commercial contacts found by Patrick Lui from the SLAC office of technology licensing.

Our work on microdrop generation is far from finished. There is ongoing work in our lab being done on the physics of operation of tubular reservoir drop ejectors by post-doctoral physicist Sewan Fan, design and experimental work on apparatus able to precisely charge control microdrops to the limits allowed by thermodynamics by physicists Martin Perl, Peter Kim, and associate engineer Howard Rogers. We have just begun collaborative research into the physics and chemistry of colliding microdrops with professor Frank Szoka of the University of San Francisco. The presentation of the results of this work along with our, at this time, limited experiments with continuous jet microdrop devices will have to await the next edition of this book.

Author

Eric R. Lee received his B.S. and M.S. degrees in electrical engineering from the University of California Berkeley and Stanford University, respectively. He has worked as an R & D engineer for medical and particle physics research groups. He is currently project manager for the microdrop particle search at Stanford Linear Accelerator Center in Menlo Park, CA.

Introduction

The use of fluid microdrops in engineering and experimental science goes back over a century to the study and use of spray generated aerosols, and the use of aerosol produced microdrops to confirm the predictions of fluid mechanics, atomic theory and chemistry. The ability to generate fluid microdrops with a predetermined size, on demand, with precisely controlled trajectories is a more recent invention dating back only a few decades. Its primary commercial and industrial use today is in the field of inkjet image printing.

Recently more exciting uses for precisely controlled inkjet generated microdrops have appeared. Due to the increased sensitivity of detectors, the need for large scale combinatorial chemistry assays using very high cost chemicals, and the need for microdispensing of small subnanoliter volumes of fluids for the making of hybrid sensors, flat screen displays, and biochips, there has been an increased interest by both industry and basic research facilities in the use of inkjet microdrops for the precision dispensing of scientific reagents for manufacturing and applied research. In addition the physical science community has increasingly been using microdrops for creating isolated microenvironments for the study of fundamental physical, optical and chemical phenomenon.

In 1992 at the Stanford Linear Accelerator Center, the research team of which I was a member had to develop devices for producing controllable streams of fluid microdrops in order to perform a search for isolated stable fractionally charged matter. In order to obtain the mass throughput and level of charge measurement accuracy we desired, the drops had to be less than 10 microns in diameter, uniform in size and produced reliably on demand over a year long continuously running experiment. The drops we needed were smaller than those produced by state of the art inkjet printers. Subsequent experiments required that we produce arrays of falling drops and drops composed of a suspension of solid meteoric material. At the time there was nothing commercially available that could generate the microdrops we required. We ultimately did successfully develop designs for drop generators and methods of formulating our own ejection compatible fluids. After we started presenting the results of our microdrop based experiments we were contacted by numerous other research groups and companies that wished to use precision generated microdrops for their projects. We discovered that in addition to its use in printing, microdrop technology was being used or proposed for use in areas as diverse as optics, drug discovery, analytical chemistry, biotechnology, and electronics manufacture.

One problem with this field as it currently exists for researchers who are not part of a large organization that has considerable internal microdrop technology expertise is the lack of easily accessible literature to guide one's initial design, prototyping and use of practical microdrop systems. The published information available for those who need to construct and operate microdrop generators is scattered throughout dozens of different journals with few if any of these papers being usable as a practical how-to guide for a person new to the field. The motivation for compiling this book occurred when we started to write short sets of operating instructions for the microdrop generators that we would periodically lend to other users. We realized that what was needed was a practical hands-on manual on the construction and use of microdrops in experimental science and manufacturing oriented towards end users who have no prior experience in generating microdrops or designing jettable fluids.

Table of Contents

CHAPTER 1

What Can You Do With a Microdrop?

Other than for printing documents, how and why has the generation of fluid microdrops, which are produced on a drop-on-demand manner, been useful to science and industry? The answer lies in the unique qualities of microdrops generated by drop-on-demand devices.

1.1 CHARACTERISTICS OF MICRODROPS

1.1.1 Size

Microdrops can be generated on demand from sizes ranging from a few microns to tenths of a millimeter. Their well-defined shape and composition, coupled with their small size, low mass, and ability to be ejected with a precise predetermined trajectory, are the enabling features for a large number of scientific and industrial applications.

As carriers for biological compounds and microorganisms, microdrops can provide a good size match for precise metering of the intended payload. In flow cytometry for instance appropriately sized microdrops have acted as carriers for individual cells for the purpose of sorting (Table 1.1).

The small mass of microdrops, comparable to some theoretically possible high mass elementary particles, has interested particle physicists as being useful for searching for exotic stable particles with high mass or fractional electric charge (Table 1.2).

The diameter of fluid microdrops, being as small as microns, has allowed levitated fluid microdrops containing fluorescent compounds to act as high Q optical resonating cavities.

In biotechnology, reagents available only at very high cost or in very limited quantities may be required to be cross-reaction tested with hundreds of thousands

Table 1.1   Size Range of Microdrop Dimensioned Objects

Object	Size in Microns
Tobacco smoke	0.25
Virus	0.1
Compacted DNA (5000 base pairs)	0.04–0.2
IC manufacturing photolithography limit	0.2
Bacteria	1
Open DNA strand (5000 base pairs)	1
Standard pigments	1–5
Inkjet pigments	0.1
Red blood cell	8
SLAC automated Millikan drops	7–20
Typical animal cell	10
Flour dust	15–20
Inkjet printer drop	18–40
Pollen	15–70
Spray can mist	1–100
Human hair diameter	100

Table 1.2   Volume and Mass of Microdrops

Drop Diameter (Microns)	Drop Volume (Liters)	Drop Mass (Grams)	Drop Mass (Gev)
0.1	5.2×10^{-19}	5.2×10^{-16}	2.9×10^{8}
0.5	6.5×10^{-17}	6.5×10^{-14}	3.7×10^{10}
1.0	5.2×10^{-16}	5.2×10^{-13}	2.9×10^{11}
5.0	6.5×10^{-14}	6.5×10^{-11}	3.7×10^{13}
10.0	5.2×10^{-13}	5.2×10^{-10}	2.9×10^{14}
25.0	8.2×10^{-12}	8.2×10^{-9}	4.6×10^{15}
50.0	6.5×10^{-11}	6.5×10^{-8}	3.7×10^{16}
100.0	5.2×10^{-10}	5.2×10^{-7}	2.9×10^{17}

of compounds. The only feasible way of performing these tests is with automated microdrop based analysis.

The use of microdrops for the combining of reagents has been proposed. The small size of the microdrops in addition to minimizing the use of reagents also minimizes the mixing time of reagents by diffusion. Automated microdrop-based titration has proved to be 10–100 times faster than conventional volumetric titrations.

For certain applications, such as the creation of aerosols for combustion, drug inhalation, or aerial dispersion, the size of the individual fluid droplets is critical for effective performance of the function of the microdrops. The creation of a mono-disperse set of drops is a far more efficient way of dispersing the fluids than the use of a random spray aerosolizer.

1.1.2   Precision Deposition Ability

Microdrop drop-on-demand devices, when engineered properly, produce drops with identical diameters and ejection speeds to within a fraction of a percent. The ability to accurately place precisely metered volumes of fluids has applications in fields from manufacturing, medical diagnostics, and metrology.

1.1.3 Isolation

Microdrops have applications where isolation of the fluid from contaminants, surfaces of containment vessels, and external disturbances is critical.

For example, the standoff distance that a microdrop generator can have from the ejection aperture to the target can be as much as centimeters for large diameter (~100 micron) microdrops. The ability to remotely place fluids in a noncontacting mode in precise locations eliminates one source of cross contamination that may result from the use of conventional fluid dispensers.

Levitated microdrops have been used as isolated self-contained reaction vessels for studies in which the walls of a solid containment vessel would interfere with the study of such phenomena as supersaturated solutions.

Levitated microdrops containing fluorescent compounds have used the continuous discontinuity-free microdrop fluid-to-air interface and the principle of total internal reflection to trap the emitted photons by having them orbit along the inner surface of the spherical microdrop.

1.1.4 High Rate Production

The maximum rate of drop production from single ejectors can be in the tens of kHz for drop-on-demand devices to the MHz range for continuous jet devices. This high rate of production combined with automated sensing and control make possible applications such as microdrop accretion of three-dimensional structures, rapid manufacture of biochip arrays, and rapid combinatorial testing of bioactive compounds for the purpose of drug discovery.

1.2 USE OF MICRODROPS IN PURE SCIENCE

Microdrop-based experiments have been performed and are still being conducted to enhance our theoretical understanding of fluid dynamics, chemistry, optics, and subatomic particle physics. Most of the basic science experimental work used microdrops as controlled isolated objects and microenvironments employing various forms of artificial levitation or other forms of motion control.[1-5] Levitation and motion control techniques used for these microdrop studies have included:

- Millikan apparatus
- Electrodynamic balances
- Quadrapole trap
- Optical levitation
- Acoustic levitation
- Microgravity
- Aerodynamic levitation

The small masses of the microdrops facilitated the use of these techniques to create isolated, physically well-defined small particle and fluid microenvironments in which to perform the desired measurements.

1.2.1 Particle Physics

The small masses of fluid microdrops along with their predictable composition and the ability to accurately measure their masses and motions allow the characterization of the forces acting on the drops. The use of microdrops as precision test objects for the study of forces on matter provided strong evidence for the quantization of electric charge and allowed the determination of the value of this fundamental unit.[6,7] Nearly a hundred years after Robert Millikan's oil drop studies determined the value of the electric charge, microdrops are again relevant to particle physics. Some speculative particle physics theories allow for the existence of stable fractionally charged particles and very massive subatomic particles.[8] Very high throughput automated Millikan microdrop measurement apparatus have been built to search for stable fractionally charged matter.[9–16] Designs have been proposed for microdrop based searches for stable massive subatomic particles. Another approach for detecting exotic particles using microdrops was to make a large-scale three-dimensional array of metastable superheated microdrops that will vaporize when impacted by a cold dark matter particle.[17,18] Laser vaporized liquid metal microdrops are being currently researched at UC Berkeley by professor Dmitry Budker's research group for use in atomic physics as a source of ultra pure contamination free vapor for spectroscopy studies. These state of the art spectroscopic systems are capable of resolving line shifts that can detect nuclear parity violations. Some of these microdrop based basic atomic and particle physics experiments conducted in the past and being presently pursued are:

- Measurement of the value of the electronic charge
- Search for fractionally charge particles
- Search for high mass elementary particles
- Detector for cold dark matter
- Spectroscopy vapor source for atomic physics

1.2.2 Fluid Dynamics

The small size of the fluid microdrops allows the study of the fluid dynamic phenomenon in which the molecular quantization of the media becomes important. This allowed studies that included:

- Stokes Law and its breakdown in the regime of very small particles. The transition between the dominance of turbulent and laminar flow for drag resistance of objects of different sizes can be directly studied with microdrops of different sizes falling in air.
- Brownian motion, which could be directly observed and correlated with droplet size and media temperature and pressure.

- Thermophoretic force measurements, which were taken on microdrops in controlled temperature gradients.

1.2.3   Physical Optics

The dimensions of fluid microdrops similar to the wavelength of light allowed studies into the nature of optical scattering in the realm where neither diffractive nor geometric optics totally dominate.[19] Other studies performed were in the areas of interferometric measurement, utilizing the internal reflectivity of the drop air interface to form an optical resonant cavity, and specialized forms of spectroscopy involving substances carried by the fluid microdrops.[20,21] Some of the studies involved:

- Mie theory scattering of light
- Single particle light scattering experiments
- Microparticle Raman spectroscopy
- Microparticle photophysics

1.2.4   Physical Chemistry

Microdroplets are a physically well-defined, easily monitored fluid system, which allow testing the correctness of theories of evaporation, condensation, behavior of supersaturated solutions, and reaction rate chemistry, in a containerless system.[5,22,23] Some of these fluid microdrops studies include:

- Evaporation rate in saturated and unsaturated environments
- Kelvin effect (effect of fluid curvature on vapor pressure)
- Knudsen evaporation (noncontinuum regime evaporation)
- Multicomponent fluid evaporation
- Hygroscopic droplet growth
- Study of supersaturated fluids
- Effect of trace additives (contaminants) on evaporation rate of fluid droplets
- Effect of gas flow on evaporation rate
- Gas–fluid chemical interaction dynamics
- Explosive boiling
- Polymerization
- Thermal mass measurement
- Rayleigh limit droplet disintegration due to charge

1.3   USE OF MICRODROPS IN APPLIED SCIENCE

1.3.1   Combinatorial Chemistry

The field of combinatorial chemistry is starting to utilize drop-on-demand inkjet ejectors to automate the mixing of reagents in different proportions. In combinatorial chemistry, one attempts to find optimal proportions of different compounds by brute force testing off all possible combinations and proportions. The small fluid volumes and rapid dispensing capabilities of drop-on-demand inkjets make possible the rapid testing of tens of thousands of mixtures in the time that it formerly took to test one new compound.[24, 25]

1.3.2 Micromixing

One of the problems that arises when making mixtures of compounds by pipetting samples of different chemicals together in a sample well is that the pipetted drops will only contact each other at their boundaries producing gradients and imperfect mixing unless mechanically stirred. Using microdrop drop-on-demand dispensers to produce the mixtures from hundreds of micron-sized drops, rather than a much smaller number of conventionally pipetted microliter-sized droplets, has the advantages that the drop boundaries are small and turbulent mixing is produced by the impact at up to 10 meters/second of the microdrops on the previously deposited liquid.

1.3.3 Automated Microtitration

Drop-on-demand microdrop dispensers combined with sensitive optical sensors can be used to build a device that can perform rapid automated titrations on very small samples of fluid. One research team at the Royal Institute of Technology in Stockholm, Sweden, demonstrated a piezoelectric drop-on-demand-based system that was capable of performing an automated tritration on a 9 nanoliter sample volume in 5 seconds.[26]

1.3.4 MALDI TOF Spectroscopy Sample Loading

*M*atrix *A*ssisted *L*aser *D*esorption *I*onization *T*ime *O*f *F*light (MALDI TOF) spectroscopy is a technique for obtaining the molecular masses of the components of an unknown compound by the ionization and time of flight measurement of a laser-vaporized sample of the unknown material that was initially deposited into a solid matrix surface. Increases in the sensitivity of the MALDI instrumentation, as well as samples of exotic materials available in only very small quantities, has placed a premium on being able to deposit nanoliter to picoliter volumes of fluid on the MALDI matrix. Microdrop ejectors are good matches to this task, as they can rapidly apply small amounts of fluids to precisely defined locations in a noncontacting mode that will not damage or contaminate the substrate.[27,28]

1.3.5 Loading and Dispensing Reagents from Microreactors

Microreactors, in the popular literature, have been called "labs on a chip." Microreactors use micromachined structures, advanced sensors, and computer control to miniaturize the chemical synthesis and analysis process to the point where a chemical process, formerly requiring a lab bench or an entire room to be performed, now can be done on a square-cm micromachined wafer.[29] The ability to synthesize hazardous or limited lifetime reactants at the immediate point of use on a real-time, as-needed basis and the ability to customize these reactants in response to external sensors are some of the operational advantages of these miniaturized chemical processing chips. There is additional motivation to reduce the volume of reactants involved in order to reduce the amount of chemical waste produced and the cost of reagents. The ordinary macroworld- to microchip-scale interface is a difficult problem with any micromachined technology. Inkjet-like fluid drop ejectors can be one solution to loading reagents into these microreactors and dispensing their final products.

1.3.6   Gas Flow Visualization

One of the difficulties of using very small microdrops in the sub-10 micron diameter range is the difficulty of precisely depositing them due to the perturbing effect of air convection. This problem of rapid coupling to local gas flow can be turned into a solution, if microdrops injected into a gas flow region are used to track the motion of the gas.

1.4 BIOTECHNOLOGY APPLICATIONS OF MICRODROPS

1.4.1   Cell Sorting

A microdrop approximating 10 to 20 microns in diameter is about the size needed to contain a single cell. Sorting systems based on microdrops have been constructed that utilize continuous jet microdrop generators. The ejected drops are electrically charged after fluorescence-based detection of the desired cell lines. Electric field plates then deflect the drops into different holding vessels. This is a mature, commercially available technology developed in the late 1960s.[30]

1.4.2   DNA Microarrays

A microarray is a two-dimensional grid of tagged DNA fragments that is used as an analytic detection method to rapidly detect genetic patterns, such as a disability causing genetic defects or a presence of disease-causing microorganisms.[31] The number of genetic patterns that can be detected increase with the number of elements in the array. However, the raw, tagged genetic material is very expensive. Therefore, the researcher is motivated to make the volume of each spot of fluid carrying genetic material markers as small as possible. The density of these arrays is that of tens of thousands to hundreds of thousands of spots per cm^2. An inkjet dispenser that shoots out microdrops is one of the three major technologies currently being used to manufacture these DNA biochip arrays.[32,33]

1.4.3   DNA Synthesis

In addition to laying out patterns of presynthesized DNA markers, jetted microdrops are being used to manufacture microarrays by direct synthesis of DNA chains. The general method is to load four microdrop ejectors with each of the nucleotide bases and deposit them on a specified reaction spot within a large array in sequence.

As each spot has a base nucleotide added and reacted, it is washed away prior to the reagent for the next base in the chain being jetted onto the spot. In this way, a DNA microarray can be built up from scratch.[32]

1.4.4 Drug Discovery

Drug discovery experiments typically require thousands to hundreds of thousands of tests involving novel biologically active agents. Microdrop dispensers that reduce the quantity of fluids used and the time it takes to do the necessary combinatorial mixing can greatly speed the testing and reduce the cost of the raw materials.[24, 25]

1.4.5 Medical Therapeutics

Inhalation of aerosolized microparticles containing therapeutic agents have been proposed as a means of delivering drugs that otherwise would have to be administered via a needle and syringe. Microdrop generators have been discussed as a means of producing the 1- to 5-micron diameter aerosolized particles needed to deliver drugs to the lungs, where they can be absorbed.

MicroFab Technologies, Inc. (1104 Summit Ave., Suite 110, Plano, TX) has published papers and has been granted patents on novel uses of microjetted fluids for biological applications, including use of ejected microdrops with tuned lasers to enhance the effectiveness of microsurgery and laser dentistry.[34]

1.5 APPLICATIONS IN MANUFACTURING AND ENGINEERING

1.5.1 Optics

The increased importance of semiconductor interfaced optical components — in particular, fiberoptics, laser diodes, imaging arrays, displays, and optical switches — has produced a need for microscale optics to effectively couple light to and from these components.

Microdrops produced from drop-on-demand ejection devices have been used to make lenses and lens arrays suitable for use in these new optical communication and imaging components.[35–37] Instead of glass, ultraviolet-cured optical plastics are used. Instead of forming optical surfaces by grinding and polishing, the microdrop-produced optics define the lens curves by control of surface tension and contact angle. Anamorphic lenses used, for instance, to correct the astigmatism in edge-emitting laser diodes can be fabricated by forming lenses from multiple merged drops laid down in asymmetric patterns. Aberration correction could be implemented by depositing optical plastics with difference indexes of refraction and dispersion over previously deposited microlenses to form the equivalent of multi-element achromats.

The specific near-term applications for these fluid-jet deposited microlenses are as collimating lenses for laser diode arrays and collimating elements to be deposited over the ends of optical fibers. Other proposed uses are large-area collimating lenses for display screens and solid-state camera photosensor arrays.

Microdrops in the 10-micron diameter range have been studied for potential use as optical resonating cavities for very low threshold lasers, hybrid bio-optical detectors, and high efficiency optical bandwidth filters.[21]

1.5.2 Droplet-Based Manufacturing

Droplet-based manufacturing is the use of inkjet techniques to accrete solid three-dimensional structures. It has also been proposed as a means of making complex composite solids, such as directionally asymmetric metal ceramic matrices that cannot be produced by conventional mixing or alloying.[38-40] The droplet can be composed of high-temperature liquid metal or hardenable polymers. Commercial products using this technique for making three-dimensional objects for industrial prototyping have been available since 1995 from companies such as Z Corporation (20 North Avenue, Burlington, MA), Solidscape Inc. (316 Daniel Webster Highway, Merrimack, NH), Objet Geometries Ltd. (Kiryat Weizmann Science Park, P.O. Box 2496, Rehovot 76124, Israel), and Sanders Design International, Inc. (Pine Valley Mill, P.O. Box 550, Wilton, NH). The size of the droplets used in researching this technique have been from 25 to 1000 microns in diameter. While the present work has concentrated on accretion of materials using microdrops, it is also possible in principle that the microdrops can carry solvents that can do subtractive synthesis.

1.5.3 Inkjet Soldering

MicroFab Technologies, Inc. has demonstrated hardware capable of depositing molten solder from inkjet ejectors onto circuit boards. The company's ejectors generated solder drop sizes as small as 25 microns in diameter. This inkjet solder technology can be used to make direct solder connections of components, vertical and horizontal conductive traces, and solder bumps for flip chip bonding.[41-44]

1.5.4 Precision Fluid Deposition

Inkjet devices rapidly place fluids, such as lubricants, where needed during manufacture and assembly with precise control over quantity and position. For example, such a device can jet microdrops of lubricants into the bearings of watches and other microgeared mechanisms without depositing excess fluids to where they would interfere with the operation of other parts of the mechanical device. Similarly thread-locking compounds can be jetted into assembly screws and pins.

1.5.5 Displays

Large area displays utilizing organic LEDs can, in principle, be fabricated with precisely deposited microdrops.[45-48] Both the light-emitting organic LEDs and the conductors can, in principle, be printed onto the substrate with inkjet devices. One form of advance display technology that is explicitly dependent upon microdrop technology for its manufacture is the Gyricon invented at Xerox that uses electro-statically charged microspheres formed from fluid microdrops that have contrasting colors on each charged hemisphere. An application of a localized electrostatic field rotates the microspheres producing local color changes to form the desired image.

1.5.6   Thin Film Coating

The method most commonly used for thin film coating of surfaces is vacuum deposition. One proposed alternative is to use microdrop deposition of fluids using contact angle and viscosity to control the film thickness.

1.5.7   Heat Radiators

One characteristic of a volume of fluid ejected into a mass of microdrops is its large increase in its surface area. A high surface-area-to-volume ratio makes for a very efficient heat radiating system. One proposed use of microdrops-based heat exchangers is for radiating away heat in spacecraft. A heat transfer fluid at a high temperature would be ejected as a volumetrically dense parallel sheet of microdrops that would be collected and recirculated. A similar structure was proposed to act as an aerobrake. The advantage is that no large, heavy, traditional, flat plate radiating surfaces are required. The fluid that removes the heat from the active components of the space ship also acts as the large surface area radiator.[49]

1.5.8   Monodisperse Aerosolizing for Combustion

The combustion rate of a volume of fluid is a strong function of the size of the aerosolized particles and the droplet size distribution. Droplet jet microdrop production allows precise control over the size of these particles and their concentration. This can be valuable for fundamental research and for future high efficiency internal combustion engines.

1.5.9   Monodisperse Aerosolizing for Dispersing Pesticides

For applications such as crop spraying, there is an optimal size for fluid droplets in order to control their rate of fall in air. A monodisperse method of forming drops is theoretically superior to using a conventional high-pressure nozzle aerosolizer.

1.5.10   Document Security

This is an ironic use of microdrop technology since one of the major facilitators of counterfeit documents has been the color inkjet printer. The real time combinatorial mixing ability of inkjet systems can be used to print difficult-to-replicate characters and logos composed of microdots, each microdot having a unique optical signature generated by cross interacting different proportionate mixtures of the printers' colorants. If spectrally nonlinearly mixing fluids are used as the colorants, the original composition of each microdot can be very difficult to reconstruct. This can be the basis of a trap-door printing scheme for printing difficult-to-counterfeit security labels.

1.5.11 Integrated Circuit (IC) Manufacturing

There are processing steps in the manufacture of integrated circuits where the uses of microdrop jetting devices have been proposed to improve upon the current state of the art.

1.5.12 IC Manufacturing — Photoresist Deposition

The deposition of photoresist is currently done by a spin-on process that throws away over 95% of the photoresist initially applied to the wafer. If a microdrop jet-based direct could apply the photoresist printing process, the cost of materials and toxic waste disposal could be significantly reduced.[50] There is also speculation that if the inkjet printing of photoresist can be done at a high enough spatial resolution, the optical lithography step can be eliminated by direct printing of the desired masking patterns.

1.5.13 IC Manufacturing — Conductor and Insulating Dielectric Deposition

Polymer-based insulating layers can in principle be applied by inkjet deposition and have the advantage over conventional processing of not requiring high temperatures. In addition, if the required spatial resolution is low, the insulating coatings can be selectively applied without requiring photolithography masking.[41-43] Similarly, liquid metal or electrically conductive solidifying fluids can be jetted and applied to form-conductive traces.

1.5.14 IC Manufacturing — Depositing Sensing and Actuating Compounds

The trend towards integrated microelectromechanical systems on a chip has produced a need to deposit sensing and actuating materials on integrated circuit chips that cannot be applied using the traditional photomasking and bulk deposition methods. Some of these compounds are composed of fragile organic molecules and would be destroyed by conventional processing. Direct microdrop jetting of these compounds into the desired chip locations during the final manufacturing step is one solution to this problem.[51]

REFERENCES

1. A. Ashkin, Applications of laser radiation pressure, *Science*, vol. 210, no. 4474, pp. 1081–1088, 1980.
2. C.A. Rey et al., Acoustic levitation techniques for containerless processing at high temperatures in space, *Metall. Trans. A*, vol. 19A, pp. 2619–2623, 1988.
3. W.B. Whitten et al., Single-molecule detection limits in levitated microdroplets, *Anal. Chem.*, vol. 63, no. 10, pp. 1027–1031, 1991.

4. Y. Ishikawa and S. Komada, Development of acoustic and electrostatic levitators for containerless protein crystallization, *Fujitsu Sci. & Tech. J.*, vol. 29 no. 4, pp. 330–338, 1993.

5. E.J. Davis, A history of single particle levitation, *Aerosol Sci. & Tech.*, vol. 26, pp. 212–254, 1997.

6. R.A. Millikan, The isolation of an ion, a precision measurement of its charge, and the correction of Stokes' law, *Phys. Rev.*, vol. 32, no. 4, pp. 349–397, 1911.

7. V.D. Hopper and T.H. Laby, The electronic charge, *R. Soc. of London Proc.*, A 178, pp. 243–272, 1941.

8. M.L. Perl, New method for searching for massive, stable, charged elementary-particles, *Phys. Rev. D*, vol. 57, no. 7, pp. 4441–4445, 1998.

9. C.L. Hodges et al., Results of a search for fractional charges on mercury drops, *Phys. Rev. Lett.*, vol. 47, no. 23, pp. 1651–1653, 1981.

10. D.C. Joyce et al., Search for fractional charges in water, *Phys. Rev. Lett.*, vol. 51, no. 9, pp. 731–734, 1983.

11. M.L. Savage et al., A search for fractional charges in native mercury, *Phys. Lett. B*, vol. 167B, no. 4, pp. 481–484, 1986.

12. C.H. Hendricks et al., Efficient bulk search for fractional charge with multiplexed Millikan chambers, *Meas. Sci. & Tech.*, vol. 5, pp. 337–347, 1994.

13. N.M. Mar, A new search for elementary particles with fractional electric charge using an improved Millikan technique, Ph.D. dissertation, Stanford University, 1996.

14. N.M. Mar et al., Improved search for elementary particles with fractional electric charge, *Phys. Rev. D*, vol. 53, no. 11, 1996.

15. V. Halyo et al., Search for free fractional electric charge elementary particles using an automated Millikan oil drop technique, *Phys. Rev. Lett.*, vol. 84, no. 12, pp. 2576–2579, 2000.

16. D. Loomba et al., A new method for searching for free fractional charge particles in bulk matter, *Rev. Sci. Inst.*, vol. 71, no.9, pp. 3409–3414, 2000.

17. J.I. Collar, Superheated microdrops as cold dark matter detectors, *Phys. Rev. D*, vol. 54, no. 2, pp. R1247–R1251, 1996.

18. J.I. Collar et al., First dark matter limits from a large-mass, low-background, super-heated droplet detector, *Phys. Rev. Lett.*, vol. 85, no. 15, pp. 3083–3086, 2000.

19. H.B. Lin, J.D. Eversole, and A.J. Campillo, Vibrating orifice droplet generator for precision optical studies, *Rev. Sci. Inst*, vol. 61, no. 3, pp. 1018–1023, 1990.

20. S. Arnold, Fluorescence spectrometer for a single electrodynamically levitated micro-particle, *Rev. Sci. Inst.*, vol. 57, no. 9, pp. 2250–2253, 1986.

21. S. Arnold, Microspheres, photonic atoms, and the physics of nothing, *Am. Scientist*, vol. 89, pp. 414–421, 2001.

22 K. Ajito, Direct structural observation of liquid molecules in single picoliter micro-droplets using near-infrared Raman microprobe spectroscopy combined with laser trapping and chemical-tomographic imaging techniques, *Thin Solid Films*, vol. 331, no. 1–2, pp. 181–188, 1998.

23. B. Kramer et al., Homogeneous nucleation rates of supercooled water measured in single levitated microdroplets, *J Chem. Phys.*, vol. 111, no. 14, pp. 6521–6527, 1999.

24. A. Schober et al., Accurate high-speed liquid handling of very small biological samples, *BioTechniques*, vol. 15, no. 2, pp. 324–329, 1993.

25. T.C. Tisone, Dispensing systems for miniaturized diagnostics, *IVD Technology*, May 1998.

26. E. Litborn et al., Nanoliter titration based on piezoelectric drop-on-demand technology and laser-induced fluorescence detection, *Anal. Chem.*, vol. 70, pp. 4847–4852, 1998.

27. G. Allmaier, Picoliter to nanoliter deposition of peptide and protein solutions for matrix-assisted laser desorption/ionization mass spectrometry, *Rapid Communications in Mass Spectrometry*, vol. 11, pp. 1567–1569, 1997.

28. J. Kling, MALDI chip shot, *Anal. Chem.*, pp. 68A–70A, Feb. 2001.

29. M. Freemantle, Downsizing chemistry, *Chem. & Eng. News*, vol. 77, no. 8, pp. 27–36, 1999.

30. H.R. Hulett et al., Cell sorting: automated separation of mammalian cells as a function of intracellular fluorescence, *Science*, vol. 166, pp. 747–749, 1969.

31. G.C. Gabriel, Microarrays: Existing and evolving genomic platforms, *Biomed. Prod.*, vol. 10, p. 26, 1999.

32. T.P. Theriault, S.C. Winder, and R.C. Gamble, Application of ink-jet printing technology to the manufacture of molecular arrays, in *DNA Microarrays — A Practical Approach*, M. Schena, Ed., Oxford University Press, Oxford, 1999, chap. 6.

33. D. Englert, Production of microarrays on porous substrates using noncontacting piezoelectric dispensing, in *Microarray Biochip Technology*, M. Schena, Ed., Eaton Publishing, Natick, MA, 2000, chap. 11.

34. D.J. Hayes, H.L. Matthews, and M.M. Judy, Method and apparatus for improving laser surgery, U.S. Patent 5,092,864, March 3, 1992.

35. W.R. Cox et al., Micro-jet printing of refractive microlenses, OSA Diffractive Optics and Micro-Optics Topical Meeting, Kailua-Kona, Hawaii, June 1998.

36. W.R. Cox et al., Micro printing of micro-optical interconnects, *Int. J. Microcircuits & Electron. Packag*, vol. 23, no. 3, pp. 346–351, 3rd Quarter, 2000.

37. W.R. Cox, C. Guan, and D.J. Hayes, Microjet printing of micro-optical interconnects and sensors, Proceedings, SPIE Photonics West 2000, paper #3952B-56, San Jose, CA, January 2000.

38. F.Q. Gao and A.A. Sonin, Precise deposition of molten microdrops: The physics of digital microfabrication, *Proc. R. Soc. of London A,* vol. 444, no.1922, pp. 533–554, 1994.

39. S. Ashley, Rapid prototyping is coming of age, *Mech. Eng.*, vol. 117, no. 7, pp. 62–68, 1995.

40. I. Austen, Molecules engineered with an inkjet printer, *The New York Times*, May 25, 2000, p. D15.

41. D.B. Wallace, Automated electronic circuit manufacturing using ink-jet technology, *Trans. of the ASME*, vol. 111, pp. 108–111, 1989.

42. D.J. Hayes, W.R. Cox, and M.E. Grove, Low-cost display assembly and interconnects using ink-jet printing technology, Proceedings Display Works '99, San Jose, Feb. 1999.

43. D.J. Hayes, D.B. Wallace, and W.R. Cox, Microjet printing of solder and polymers for multi-chip modules and chip-scale packages, Proceedings IMAPS '99, Chicago, Oct., 1999.

44. M. Orme et al., Electrostatic charging and deflection of nonconventional droplet streams formed from capillary stream breakup, *Physics of Fluids*, vol. 12, no. 9, pp. 2224–2235, 2000.

45. J. Bharathan and Y. Yang, Polymer electroluminescent devices processed by inkjet printing: I. Polymer light-emitting logo, *Appl. Phys. Lett.*, vol. 72, no. 21, pp. 2660–2662, 1998.

46. S. Chang et al., Dual-color polymer light-emitting pixels processed by hybrid inkjet printing, *Appl. Phys. Lett.*, vol. 73, no. 18, pp. 2561–2563, 1998.

47. T.R. Hebner and J.C. Strum, Local tuning of organic light-emitting diode color by dye droplet application, *Appl. Phys. Lett.*, vol. 73, no. 13, pp. 1775–1777, 1998.

48. M. Grove et al., Color flat panel manufacturing using ink jet technology, Proceedings Display Works '99, San Jose, Feb. 1999.
49. E.P. Muntz and M. Dixon, Applications to space operations of free-flying, controlled streams of liquids, *J. Spacecraft*, vol. 23, no. 4, pp. 411–419, 1986.
50. G. Percin, T. Lundgren, and B.T. Khuri-Yakub, Controlled ink-jet printing and deposition of organic polymers and solid particles, *Appl. Phys. Lett.*, vol. 73, no. 16, pp. 2375–2377, 1998.
51. R. Rinaldi et al., Photodetectors fabricated from a self-assembly of a deoxyguanosine derivative, *Appl. Phys. Lett.*, vol. 78, no. 22, pp. 3541–3543, 2001.

CHAPTER 2

Methods of Generating Monodisperse Microdrops

Robert Millikan performed his famous levitated oil drop determination of the value of the electric charge in the early 1900s with fluid microdrops made by aerosolizing low-viscosity watch oil. He used a spray atomizer, similar to that used for dispensing perfume. Then, from the different-sized atomized drops, he selected the ones with the diameters appropriate for his measurements.

For applications requiring greater drop production precision and efficiency, the generation of uniformly sized microdrops with well-defined trajectories is a far more desirable drop-generating process than the production of a random aerosol containing drops in the general size range of the microdrops one wishes to use. In general, to accomplish this goal of ejecting monodisperse microdrops, one needs the ability to produce high-speed fluid jets of approximately the diameter of the drops one wishes to generate and then to control the behavior of the jets precisely enough to cause them to consistently condense into uniformly sized drops.

From common experience, simply pressurizing a fluid and letting it seep out of a small hole will result in an adhering fluid mass that will break off when its weight exceeds the surface tension forces holding it onto the ejection aperture. This is the mechanism by which millimeter-scale diameter droplets are produced from a leaky kitchen faucet. In contrast, in order to produce the micron-scale diameter fluid jets needed to make the microdrops used for instance in inkjet printers, the fluid pushed out of the ejection aperture hole must be traveling fast enough, with enough kinetic energy per unit volume, that it can overcome the interfacial energy attracting it to the surfaces of the ejection aperture and have sufficient additional kinetic energy to create the increased surface area per unit volume required to form a fluid microjet. The minimum fluid jet speeds required are on the order of 1 to 10 meters per second.

Given that a fluid jet of some kind is required, there are two principal ways to form microdrops from this starting point. Historically the first method used was to break up a continuously flowing fluid jet by driving the fluid with a source of acoustic energy in order to form standing wave nodes along its length. The nodes along the jet would condense into discrete microdrops. The other method of forming discrete monodisperse microdrops from a fluid jet is to make short duration fluid jets instead of breaking up a continuous jet, with each short duration jet condensing into a single microdrop of the desired diameter. This is the drop-on-demand method, as opposed to the previously described continuous jet technique. The drop-on-demand method is the one used for the majority of commercial inkjet printers.

The continuous stream method is capable of producing drops at MHz rates — that is two orders of magnitudes faster than the best drop-on-demand devices — but has the practical disadvantage of higher hydraulic complexity and much higher minimum operating fluid volumes than drop-on-demand devices. The minimum operating rate of these continuous stream drop generators is in the tens to hundreds of kHz range, which is in fact higher than the maximum operating frequencies of many drop-on-demand devices. Continuous jet microdrop generators have been successfully used in the printing industry in scientific applications for cell sorting and monodisperse aerosol generation. Continuous jet microdrop generators are, in general, unsuitable for applications in which small quantities of high value fluids must be microdispensed.

Due to its greater ease of use and lower cost of hardware and much smaller minimum fluid operating volumes, most recent technological research into micro-drop generation has been in the area of advanced drop-on-demand devices. In order to make up for the lower maximum rate of drop production, drop-on-demand technology utilizes miniaturization and massive paralleling of ejectors into a common functional unit. State of the art inkjet printers for instance have hundreds of parallel, independently operating microdrop ejectors integrated into each printhead. What is needed for conventional drop-on-demand operation is a low fluid impedance nozzle of approximately the diameter as the drop one wishes to eject and some kind of controllable actuator that can generate microsecond scale pressure impulses in the fluid. The principle methods of actuating drop-on-demand devices in commercialized hardware have been with piezoelectric elements, and thermally generated gas bubbles created by resistive heating elements in contact with the working fluid. Electrostatic actuation, pneumatic, inertial, thermal bimorph plates, high voltage, and spark actu-ation have also been implemented in experimental devices.

Having a constant ejection pressure applied by an external pump and implement-ing high-speed valving is another approach to drop-on-demand. Electro-rheological inkjets implement this high-speed fluid switching by the use of fluids that can rapidly change viscosity and transition from Newtonian to plastic flow characteristics in response to a high-speed, switched electric field. There is a thermally valved version of the electrohydrodynamic inkjet in which there is a constantly applied electric field that is insufficient to draw out a fluid jet until the viscosity is reduced by localized heating via a miniature pulsed resistive heating element. For making large drops that are close to a mm in diameter, conventional high-speed electromechanical valves have been used.

Two unique drop-on-demand methods can produce small-diameter fluid jets without requiring a nozzle with a diameter of the size of the microdrop. Electrohydrodynamic inkjets use direct electrostatic attraction to pull a fluid jet from the end of a capillary with the small-diameter jet emerging from the tip of a Taylor cone. Focused ultrasound inkjets use focused ultrasonic radiation pressure from lensed transducers to eject a drop from a fluid-air interface.

The electro-rheological and focused ultrasonic microdrop generation methods are relatively new developments compared with the more conventional pressure impulse drop-on-demand techniques. The electrohydrodynamic and ultrasonic methods have the advantage of being far more immune to stoppages due to clogged ejection aperture holes due to their ability to generate small drops from much larger diameter fluid ejection aperture holes. Some disadvantages of these lesser-used techniques is that the electro-rheological and electrohydrodynamic fluid jetting techniques place far tighter constraints on the rheological and electrical properties of what fluids can be used than conventional pressure impulse microdrop ejectors. Thermal bubble and spark actuated inkjet devices necessarily disturb the chemical integrity of its working fluid. For experimental science applications, the piezoelectrically driven drop-on-demand ejector, the various forms of the TopSpot® drop ejector, and the continuous jet ejector have been proven to be the most accommodating of a wide variety of fluids and the least disruptive of their operating fluids and fluid payloads during the process of droplet ejection. Recent research has indicated that focused beam ultrasound drop ejection devices may be similarly compatible with fragile microdrop payloads.

2.1 ACOUSTICALLY DISRUPTED CONTINUOUS FLUID JET

Also called the continuous inkjet, a continuous fluid jet is produced by pressurizing the fluid reservoir, which causes the fluid to be jetted out as a continuous stream of approximately the diameter of the ejection aperture nozzle (Figure 2.1). The ejector is excited with a CW acoustic waveform that causes instability and standing waves on the fluid stream as it emerges from the nozzle orifice hole. The standing wave maxima along the fluid stream then form into individual drops.[1-5] The diameter of the drops produced by the fluid stream is approximately twice the diameter of the ejection aperture. Depending upon the rheology of the fluid, the drop production rate and the size of the drops can be controllably altered to a certain extent, even with a fixed aperture, by changing the velocity of ejection of the jet and the frequency of the driving signal. For a given drive frequency, the velocity of the jet determines the rate of drop production. The vibrations induced into the fluid jet can be from a piezoelectric transducer, an electromagnetic transducer, or a compressed gas-driven mechanical resonator. The jet ejection speed referenced in the literature ranged from 2 to 50 meters/second. The minimum jet velocity is given by the requirement that the kinetic energy imparted to the fluid be sufficient to create the surface energy of the jet. Operating pressures are in the 5 to 50 psi range. The frequency can be varied to produce nodes at different intervals along the jet in order to vary, within limits, the droplet diameters. The limit on how small a series

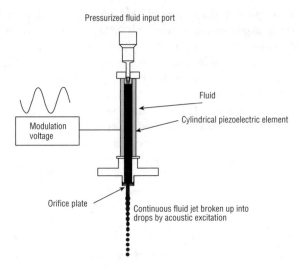

Figure 2.1   Continuous fluid jet microdrop generator. Monodisperse microdrops are
formed from a continuous fluid jet by the formation of nodes along the fluid jet
from an externally impressed source of acoustic energy.

of drops can be created is that the fluid jet is stable against perturbations of
wavelength less than three times the jet diameter. The maximum diameter of the
drops producible for a given jet diameter is limited by how long the distance can
be between nodes. This distance is constrained by satellite drop production if the
distance between the nodes is so long that harmonics can start to produce secondary
nodal points. Schneider and Hendricks[1] observed stable drop break up with nodal
intervals along the fluid stream of between 3.5 to 7 times the jet diameter. Using a
fixed diameter fluid jet, a variation by a factor of two in drop diameter has been
observed to be achievable.

    Control of the microdroplet impact points for the purpose of image printing
using this type of ejector is typically done by electrically charging the drops and
then using deflection electrodes to direct drops either into the targeted regions or
into a droplet catcher for recycling unused droplets. Control of trajectory can be
implemented by either modulating the potential between external deflection electrodes or by modulating the induced charges on the drops. This technique is the
drop-generation method used in commercial cell sorters.

2.2 THERMAL INKJET (BUBBLE JET)

    Thermal inkjet, sometimes referred to as bubble jet, is a drop-on-demand tech-
nology that uses electrical pulses applied to heating elements in contact with the
fluid near the ejection aperture nozzle in order to vaporize a small amount of liquid
to produce pressure impulses by the formation and collapse of gas bubbles[6-9] (Figure
2.2). The conversion of the electrical drive pulses to localized heating of the fluid
is mediated by thin film resistors in intimate contact with the fluid. No direct

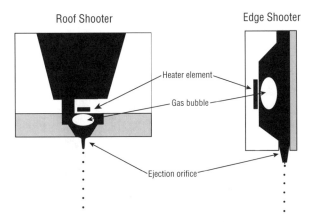

Figure 2.2 **Thermal inkjet.** Thermal inkjets, also known as bubble jets, use gas-bubble generation by localized contact heating to generate the pressure pulse needed to actuate a drop-on-demand microdrop ejector. The two forms in common use are classified by their fabrication technology as roof shooters and edge shooters. Roof shooters are fabricated by bonding an ejection orifice plate structure over the top of a wafer on which the fluid flow and heating elements are fabricated. Edge shooters, in contrast, form their ejection apertures from channels etched longitudinally into the wafer.

electrical contact with the fluid itself is needed. This pressure pulse is used to eject a jet of fluid from a small orifice that, given correct drive levels, will form into a single drop. This technique has the advantage of ease of integration into a dense print array inkjet print head, since the drive mechanism is simply a resistor placed in contact with the fluid to be ejected. The drawbacks are the requirement of a nonkoagating (thermal reactant forming) fluid, the lack of flexibility in tailoring the rise and fall time of the pressure pulse for optimizing control over the ejected fluid jet, and most importantly for scientific applications, the local chemical reactions that will take place during each vaporization and cooling cycle that can change the chemical composition of the fluid over time. On balance though, for commercial inkjet image printers, the advantages vastly outweigh the disadvantages. Hewlett Packard and Canon have manufactured highly successful lines of inkjet printers that have used this technology for over a decade.

2.3 PIEZOELECTRIC DIRECT PRESSURE PULSE

Piezoelectrically driven drop-on-demand devices operate similarly to bubble-jet microdrop ejectors except that a piezoelectric element is used to change the volume of the ink reservoir in order to produce the fluid ejection and retraction pressure pulse. Epson printers use this technology. The advantage of piezoelectric actuation is that the pressure pulse rise and fall times can be tailored to optimize monodisperse satellite free drop production and dynamically alter the diameter of the ejected drops. Also, the pressure pulse is generated in a way that does not chemically alter the composition of the fluid like the bubble-jet technique. The drawback to the piezo

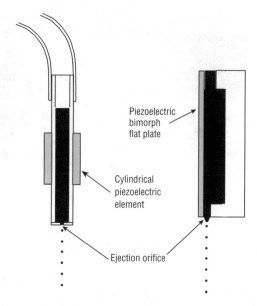

Figure 2.3   Piezoelectrically driven drop-on-demand microdrop ejectors. The two most common drive configurations likely to be encountered in nonimage printing applications are the externally actuated squeeze mode tubular reservoir drop ejector and the planar flat flex plate actuated ejector. The tubular design is easier to fabricate by hand and service but is inferior to the flat plate design in the ability to be used as elements in miniaturized close packed arrays.

technology over thermal inkjets is that the custom fabrication of an array of micro-machined piezoelectrically actuated drop generators with independently settable drive levels for each channel requires far more complex micromachining processes, since piezoelectric materials must be integrated into each ejector element in a manner that is compatible with the fluid channel etching and orifice hole formation processes.

The two major geometries used in stand-alone microdrop ejectors are the externally excited squeeze mode tubular reservoir drop generator invented in 1974 by Steven Zoltan of the Clevite Corporation[10–14] and the various flat-drive plate actuator designs[15–19] (Figure 2.3).

The tubular reservoir design, particularly when constructed from glass, has the advantages of a chemically inert fluid contact environment, ease of inspection during filling and cleaning, and ease of manufacture and handling using relatively low-cost equipment.

The flat-drive plate design, however, is better suited to be manufactured in miniaturized form by integrated circuit fabrication processes to be used as individual elements that are close-packed into parallel ejector arrays. Other flat-plate drive configuration variants that have been used in commercial print heads to actuate microdrop ejection are push-mode and shear-mode designs. Push-mode drop ejectors use piezoelectric or inert rods to transfer a mechanical impulse to an ink chamber through a flexible membrane in contact with the ink. Shear-mode designs utilize piezoelectric elements in which the direction of polarization of the drive element is nonparallel to that of the applied electric field. Shear-mode designs cause a single-element piezoelectric material to bend rather than to simply change dimension in the direction of the applied electric field without requiring a dual material actuation plate. Bend deformation in general produces a greater cavity volume change for a given electric field change than conventional actuation geometries and can simplify the manufacturing process for making high density ejector arrays.

Flat plate ejectors have been constructed in which thermal bimorphs and electrostatic actuation were used to produce the displacement needed to eject a fluid jet. Both flat plate and Zoltan style inkjet heads are commercially available as scientific/industrial stand-alone fluid ejectors for nonimage printing applications. Published literature has indicated that piezoelectric drop-on-demand microdrop ejectors can deliver intact biological payloads such as live microorganisms and DNA.[20–21]

2.4 FOCUSED ACOUSTIC BEAM EJECTION

The use of focused ultrasound beams to actuate microdrop ejection is a recently developed method invented at Xerox PARC, which holds over a hundred patents relating to this technology for producing drops on demand. This method works by focusing an ultrasound beam with an acoustic lens onto the surface of a fluid meniscus, using the acoustic pressure transient generated by the focused tone burst to eject a fluid jet[22–26] (Figure 2.4). This technique has the very large practical advantage of being potentially immune to particulate jamming of the ejection aperture since the ejection aperture is simply a region defined over a large, exposed fluid surface by the diameter of the focal spot. This can allow the reliable ejection of fluids that would otherwise clog small apertures. An enclosed fluid aperture region, though, has been reported as still being necessary to suppress fluid agitation that would destabilize the lens-to-surface distance. Another potential operational advantage is the ability to vary the size of the ejected microdrops dynamically without

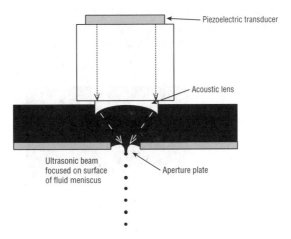

Figure 2.4   Focused acoustic beam microdrop ejector. A high frequency gated continuous wave ultrasound source is focused onto the surface of a fluid reservoir. The acoustic radiation pressure is used to produce a localized jet that forms into a microdrop.

changing any of the hardware by shifting the fluid to transducer distance in order to vary the focal spot diameter on the surface of the fluid. The ejection energy applied per drop for focused acoustic beam ejection was reported to be about 25 times that needed for a thermal inkjet.[27] This required ultrasonic ejection energy, however, does not appear to adversely affect fragile payloads such as living cells, proteins and DNA as proven by the acoustic beam microdrop ejection systems intended for biotechnology applications that have been successfully laboratory tested and are being developed for commercial use by Picoliter Inc. (231 South Whisman Road, Suite A, Mountain View, CA).

2.5 LIQUID SPARK INKJET

Developed by Olivetti in 1986,[28] this technique is similar in certain ways to the bubble jet, but the gas bubble is produced not by a resistor but by an electrical spark discharge generated by high voltage electrodes directly in contact with the fluid (Figure 2.5). One electrode is in contact with the bulk of the fluid in the reservoir. The other electrode is external to the nozzle, such that the current path to the bulk of the fluid must pass through the ejection aperture hole. The high current density produced by the concentration of current in the fluid cylinder in the nozzle vaporizes the fluid in the center of the cylindrical nozzle. This expanding vapor bubble accelerates the fluid in the front of the nozzle into a fluid jet. The inventors claim very high reliability due to the high drive voltages, in the thousand-volt range, which can arc through and break up solids that may clog the ejection aperture. One major practical difficulty in integrating this technology into inkjet array printers is the requirement for thousand-volt drive pulses. The drop ejector described by E. Manini and A. Scardovi in 1987[28] produced drops with fluid volumes of 400 picoliters, roughly equivalent to a 100-micron diameter drop. The authors did not mention how scalable in drop-size production this technology is. For scientific applications, this shares the potential problem with the bubble jet in that the ejection mechanism necessarily disturbs the chemical composition of the fluid. Given the use of high-voltage arcs to generate the vapor bubble, ablation of the aperture is another potential problem.

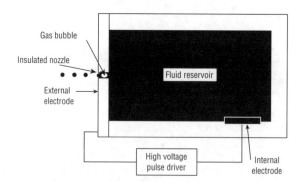

Figure 2.5   Liquid spark inkjet. A liquid spark inkjet operates similarly in principle to a thermal inkjet but generates its gas bubble with a high voltage electric arc as opposed to a resistive contact heating element.

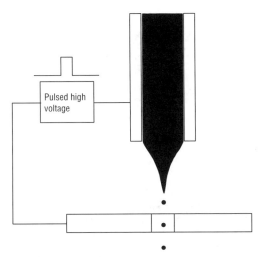

Figure 2.6   Electrohydrodynamic inkjet. An electrohydrodynamic inkjet uses a high electric
field on the order of a kilovolt per mm to pull fluid from a specially shaped and
pressurized capillary tube. Under most operating conditions a chaotic fluid spray
results. However under special conditions of proper fluid rheology, capillary geometry and precise control of the electric field monodisperse drops can be generated.

2.6 ELECTROHYDRODYNAMIC INKJET

The drawing of fluid from a capillary nozzle with an applied electric field has
been known since turn-of-the-century experimental physics. There have been many
attempts in the past few decades to adapt this technique for general purpose inkjet
printing.[4,29–34] The electrohydrodynamic inkjet functions by pressurizing the fluid so
that it forms a convex meniscus (Figure 2.6). An electric field is applied that draws
out this convex meniscus into a sharp cone. When the electric field strength is high
enough to overcome the meniscus surface tension, the fluid can break free. Depending upon the static biasing field and the duration and amplitude of the ejection pulse,
this technique can be used to produce a wide-angle spray, a continuous stream, or
under special conditions, discrete mono-disperse drops.[35–39] Due to the way the
electric field draws the meniscus out into a sharp cone, this type of inkjet is capable
of producing drops much smaller than the fluid aperture hole diameter. Mutoh in
1980[29] reported drop sizes of about an order of magnitude smaller than the aperture
diameters used. The field strengths needed to initiate drop ejection using this technique is about 1000 volts per millimeter. Typical electrode gaps were 0.5 mm. Pulse
widths were in the tens of microseconds to a few millisecond range. Choi and Lee
in 1992[30] documented the electrical characteristics of the fluid that would be suitable
for ejection by this method. They gave high dielectric constants and conductivities
in the 10^{-13} to 10^{-4} S/cm range as being necessary properties for fluids that are to
be ejected into microdrops using this technique. Low surface tension is important
because the electric field needed to eject drops is directly proportional to the fluid
surface tension.

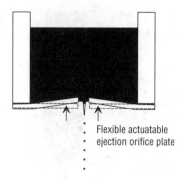

Flexible actuatable
ejection orifice plate

Figure 2.7   Flextensional aperture plate inkjet. This is a design for a mechanically actuated microdrop ejector that combines the ejection orifice plate with the mechanical actuator. Due to the integration of these two functions into one compact structure, it has the potential for making very high-density arrays of microdrop ejectors.

2.7 FLEXTENSIONAL APERTURE PLATE INKJET

The flextensional aperture plate inkjet is unique in its combining of the ejection aperture plate and the actuating mechanism into a single structure (Figure 2.7). The advantage of this technology is its potential for making highly spatially dense two-dimensional arrays of ejectors using relatively simple microstructures. The actuation mechanism for the flexible orifice plate can be from thin film piezoelectric material deposited over the orifice plate, thermal bimorph thin films, or from the electrostatic attraction of the aperture plate to an inner electrode.[40]

2.8 ELECTRO-RHEOLOGICAL FLUID INKJET

This type of drop-on-demand microdrop ejector utilizes a fluid that, under a high electric field, transitions from a Newtonian phase to a fluid with plastic flow characteristics having a high enough slippage threshold relative to the constant applied pressure that no fluid displacement occurs. Pulsing off the electric field allows the fluid to form a momentary jet. Since the motive power is provided by the external constant applied pressure and the switching is accomplished by a pair of opposed electrodes, it is very simple to fabricate high-density printing arrays using this technology[36] (Figure 2.8). The primary engineering disadvantages are the highly specialized fluid needed to make this technique work and the high electric fields on the order of kilovolts per mm needed to implement this type of fluid property modulation.

2.9 LIQUID INK FAULT TOLERANT (LIFT) PROCESS

The liquid ink fault tolerant (LIFT) process inkjet is a variation on the electro-hydrodynamic inkjet technique in which a constant subejection threshold electric

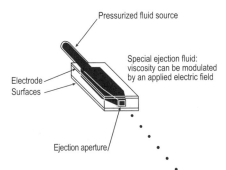

Figure 2.8   Electro-rheological inkjet. An electro-rheological inkjet operates with its fluid under constant externally supplied pressure. The fluid is designed such that under an applied electric field, it enters a plastic phase in which no flow takes place. When this field is released, the fluid has a viscosity low enough to form a high-speed jet and eject a fluid microdrop.

field is applied and an ink is used where the surface tension is a strong function of temperature (Figure 2.9). Heating elements applied to the ink channels then can switch on and off the ejection of the drops.[26] The advantage claimed for this technique over a straight thermal inkjet is that the energy per drop ejected is about 1% of that required for bubble-jet drop ejection. This reduced heat load makes a large, high-packing density print array easier to engineer. The principal disadvantages are the requirement for the use of high-voltage electrode biasing between the ejector and the disposition surface and the highly specialized fluid properties needed for the ink.

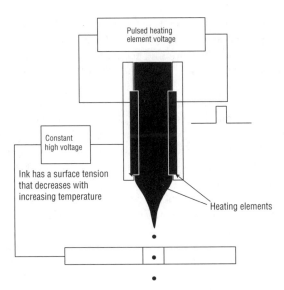

Figure 2.9   Liquid ink fault tolerant (LIFT) process inkjet. LIFT is a variation on the electro-rheological inkjet. A constant subejection threshold electric field is applied to the fluid meniscus. The fluid however is designed such that if subjected to a temperature increase, it has a reduction in surface tension sufficient for the electric field to draw out a fluid jet.

2.10 TOPSPOT® MICRODROP EJECTOR

This is a method of microdrop generation developed specifically for the making of biochip microarrays.[41–43] The principle of operation of the TopSpot method is to use capillary action filling of a fluid reservoir the volume of the desired microdrop and then forcing out this fluid by pneumatic action, elastomer displacement of the fluid, or inertial ejection. Since the printing of a biochip microarray involves the deposition of an identical grid of microdrops onto each chip, the print array does not require individual addressing of each fluid channel. Each fluid channel in a TopSpot print array can then consist only of an open top main reservoir, a narrow fill channel, and a microdrop reservoir. A common actuation mechanism can be used to eject the fluid from the microdrop reservoir from multiple channels (Figure 2.10). Since the entire volume of the fluid in the microdrop reservoir is ejected during each operating cycle, tuning of the actuation force to the fluid rheological properties to avoid satellite and spray formation is claimed to be unnecessary. One limitation of this technology is that the ejection of small drops less than 50 microns in diameter may not be practical due to the surface contact forces relative to the surface area and mass of the drops precluding easy dislodging of the fluid from the drop nozzle reservoir. This microdrop spotting device is available commercially with up to 96 channels per print head through HSG-IMIT in Germany (HSG-IMIT, Wilhelm-Schickard Srt 10, D-78052 Villing-Schwenningen, Germany).

2.11 SUMMARY AND RECOMMENDATIONS

There are many microdrop generating technologies intended for image printing that have been demonstrated in the laboratory and in commercially available products. Not all of the image printing techniques are applicable to microdrop ejection for experimental science. Table 2.1 compares the properties of several ejection technologies.

Ejection of fluids for experimental science has the additional requirement that that the microdrop ejection process not alter the chemical or physical properties of the liquid or its payload, which may in some cases be live organisms or fragile complex organic molecules. Ejection processes that apply large amounts of energy into small regions of the fluid — such as the liquid spark inkjet, bubble jet and focused ultrasonic inkjet ejectors — depending upon the design implementation of the ejector have the potential of causing serious alterations of the fluid chemistry and possible disruption of living organisms and organic macromolecules. Thermal bubble-jet drop ejectors require about a hundred times the input energy per drop ejected than comparable piezoelectrically driven drop ejectors. The focused acoustic beam drop ejector requires about an order of magnitude more applied energy to the fluid than the thermal bubble-jet drop ejector. In contrast, the TopSpot microdrop ejection technology was designed from the start not as an image printhead but as an ejection mechanism for biomolecules and consequently has a relatively benign ejection mechanism using inertial forces or direct volume displacement by a chemically inert elastomer to release a microdrop volume from a nozzle reservoir. Actual

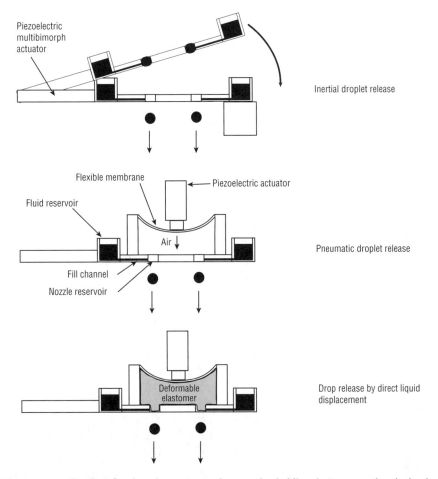

Figure 2.10   TopSpot® microdrop generation method. Microdrop generation is implemented by accumulating fluid through capillary action to fill in a micromachined nozzle reservoir, and then expelling this fluid volume. Successful methods of breaking this volume of fluid loose in the nozzle reservoir include: (a) inertial ejection in which the entire assembly is accelerated by a piezoelectric lever, (b) pneumatic ejection in which the volume of air above the nozzle reservoir is compressed by a piezoelectric actuator, and (c) direct displacement of the liquid out of the nozzle reservoir by volumetric displacement by an inert deformable elastomer driven by a piezoelectric actuator.

testing of the hardware with the intended fluid is needed though before one can establish whether the drop ejection process has any adverse effects upon the microdrop payload. Despite the high fluid accelerations and high local fluid shear rates piezoelectric ejectors have been tested as being able to eject and deposit intact live organisms and DNA.[20,21] Similarly, tests conducted by Picoliter Inc. have established that their designs for focused acoustic beam drop ejection devices can be used for the precision deposition of live cells, DNA, and proteins.

The surfaces of the drop ejector that contact the fluid should be nonreactive with the fluid. This requirement is difficult to meet with drop ejectors such as the liquid

Table 2.1 Comparison of Different Microdrop Ejection Technologies

	Fabrication Cost[a]	Cost of Drive Electronics[b]	Wide Fluid Compatibility[c]	Nondegrading of Fluids[d]	Close Packed Arrays[e]	Minimum Fluid Volume[f]	Resistant to Clogging[g]
Continuous inkjet		Low			−	High	
Thermal inkjet		Low		−	+		
Tubular piezo inkjet	Low	Low			−	Low	
Planar piezo inkjet		Low					
Flextensional inkjet		Low					
Acoustic beam inkjet		High			+		+
Liquid spark inkjet		High		−			+
Electrohydrodynamic inkjet		High	−				+
Electro-rheological inkjet		High	−		+		+
LIFT inkjet		High	−				
TopSpot drop ejector		Low	+	+		Low	

[a] The *fabrication cost* comparison is for the minimum cost procedure to produce a single microdrop ejector suitable for use in experimental science applications. Some designs are classified as expensive because they require complex microfabrication or expensive components. The tubular reservoir piezo drop ejector is bay far the least expensive to produce in small quantities as it can be made from a heated and hand ground tip glass pipette with a glued on actuator.

[b] *Cost of drive electronics* for different technology varies because of the different voltage levels required to actuate the drop ejector or the complexity of the waveform needed. Certain designs such as the liquid spark inkjet require an actuation pulse in the thousands of volts range to eject a drop while a well designed thermal inkjet or piezoelectrically actuated drop ejector may require only a few tens of volts.

[c] *Fluid compatibility* is a measure of how restrictive the technology is to the chemical, rheological and electrical properties the fluid must possess in order to be usable in that technology's microdrop ejectors.

[d] *Nondegrading of fluid* is a reference as to whether the technique for fluid ejection causes alterations in the fluid's chemical or physical properties.

[e] *Close packed arrays* refers to the relative suitability of the drop ejector technology to the making of high spatial density ejector arrays.

[f] *Minimum fluid volume* evaluation is a relative measure of whether the technology is suited to the handling and ejection of very small volumes of rare or high cost reagents. While in principle any drop ejector type can be back filled with a fluid immiscible with the desired ejection fluid this can lead to many practical problems with reagent contamination. The relative merits of each technique for operating with very small amounts of operating fluid is based upon the use of a uniform fluid mixture. The reason for the very low operating volume of the tubular piezo inkjet is its ability to have its fluid reservoir fabricated from narrow bore glass capillary pipettes and able to eject fluid even when only the tip of the tubular reservoir is filled. Continuous stream inkjets have large fluid operating volumes because in practice they have a startup cycle in which large amounts of fluids are jetted before stable operation is achieved.

[g] All drop ejectors have a nozzle-like region where fluid must pass to be ejected as microdrops, but some drop-ejector designs are able to eject microdrops with nozzle diameters far larger than the diameter of the ejected microdrop. These designs are far less likely to experience stoppages due to clogged nozzles. A clogged nozzle is the principle failure mode of microdrop ejectors. One design, the liquid spark inkjet, uses a high-voltage discharge passing through the nozzle face to vaporize fluid to the rear of the nozzle to eject a microdrop. This design is claimed to have a self-clearing operating cycle for stoppages resulting from deposits in the ejection nozzle.

spark inkjet and electrohydrodynamic inkjet, which require that conductive electrodes be in contact with the fluid, and much more easily met with pressure-actuated ejectors such as the TopSpot ejector and tubular reservoir squeeze mode piezoelectric drop ejectors.

The costs of construction and operation vary by large amounts for the different designs. Some designs, such as the tubular reservoir piezoelectrically driven drop ejector, can be constructed with common hand tools and inexpensive commercially available off-the-shelf parts. A side-shooter thermal inkjet requires sophisticated and expensive fabrication technology that is cost effective for large-run commercial production, but not for making single custom units for scientific experimentation. The cost and operational safety hazards associated with the drive electronics is also an issue. Some types of fluid drop ejectors require the high-speed switching of thousands of volts, while others, such as well-engineered piezoelectrically driven drop ejectors, can operate with pulses of a few tens of volts in amplitude.

Fluid compatibility is a big issue in the design of drop ejection systems. All ejection technologies have limits on the viscosity and surface tension of the fluids that can be ejected as microdrops. The electro-rheological inkjet and LIFT process inkjet require very special properties of their operating fluids that may preclude the use of these types of devices for the ejection of most scientifically relevant fluids and payloads. The continuous jet and pressure impulse actuated drop-on-demand ejectors have the widest tolerance range of fluid types. In fact, liquid metal has been ejected from specially designed piezoelectrically driven tubular reservoir drop ejectors.

Historically, the most common drop ejector used in laboratory science is the continuous stream droplet generator. The most commonly used drop-on-demand ejector in laboratory science is the piezoelectrically driven tubular reservoir type ejector. Both types can be constructed from low-cost, off-the-shelf components without the requirement for custom microfabrication. Fluid handling and routine maintenance are easier for transparent glass-tube-based droplet ejectors than for those made from micromachined silicon or metal. In our experiments at the Stanford Linear Accelerator Center (SLAC), we initially used a modified aluminum-and-stainless-steel-combination continuous jet and drop-on-demand microdrop ejector. After much experimental work with different microdrop generators, we eventually ended up designing and using various glass reservoir tube variations on Zoltan's piezoelectrically actuated tubular ejector design. This type of drop generator was adopted primarily for ease of maintenance, wide range of drop ejector compatible fluids, and low cost of fabrication and support electronics.

REFERENCES

1. J.M. Schneider and C.D. Hendricks, Source of uniform-sized liquid droplets, *Rev. Sci. Instrum.*, vol. 35, no. 10, pp. 1349–1350, 1964.
2. R.N. Berglund and B.Y.N. Liu, Generation of monodisperse aerosol standards, *Environ. Sci. & Tech.*, vol. 7, no. 2, pp. 147–153, 1973.

3. J. Dressler, Two-dimensional, High Flow, Precisely Controlled Monodisperse Drop Source, Gov. Doc. WL-TR-93–2049, Aero Propulsion and Power Directorate, Wright Laboratory, Air Force Material Command, 1993.

4. J. Heinzl and C.H. Hertz, Ink-jet printing, *Adv. Electron. & Electron Phys.*, vol. 65, pp. 91–171, 1985.

5. H.B. Lin, J.D. Eversole, and A.J. Campillo, Vibrating orifice droplet generator for precision optical studies, *R. Sci. Instrum.*, vol. 61, no. 3, pp. 1018–1023, 1990.

6. T. Pan, Monolithic thermal ink jet printhead with integral nozzle and ink feed, U.S. Patent 4,894,664, 1990.

7. J.D. Meyer, Thermal ink jet: Current status and future prospects, *Color Hard Copy and Graphic Arts III*, SPIE, vol. 2171, pp. 88–91, 1994.

8. J. Chen and K.D. Wise, A high-resolution silicon monolithic nozzle array for inkjet printing, *IEEE Transactions on Electron Devices*, vol. 44, no. 9, pp. 1401–1409, 1997.

9. T. Kraemer, Printers enter the jet age, *Inven. & Tech.*, Spring 2001, pp. 18–27, 2001.

10. S.L. Zoltan (Clevite Corp.), Pulse droplet ejection system, U.S. Patent 3,683,212, 1972.

11. S.L. Zoltan (Gould Inc.), Pulse droplet ejection system, U.S. Patent 3,857,049, 1974.

12. D.B. Bogy and F.E. Talke, Experimental and theoretical study of wave propagation phenomena in drop-on-demand ink jet devices, *IBM J. Res. & Dev.*, vol. 28, no. 3, pp. 314–321, 1984.

13. J.F. Dijksman, Hydrodynamics of small tubular pumps, *J. Fluid Mech.*, vol. 139, pp. 173–191, 1984.

14. G.L Switzer, A versatile system for stable generation of uniform droplets, *R. Sci. Instrum.*, vol. 62, no. 11, pp. 2765–2771, 1991.

15. E. Stemme and S.G. Larsson, The piezoelectric capillary injector — A new hydro-dynamic method for dot pattern generation, *IEEE Transactions on Electron Devices*, vol. 20, no. 1, pp. 14–19, 1973.

16. E.L. Kyser, L.F. Collins, and N. Herbert, Design of an impulse ink jet, *J. Appl. Photographic Eng.*, vol. 7, no. 3, pp. 73–79, 1981.

17. N. Takada et al., The effect of head dimensions on drop formation in drop-on-demand ink-jet printing, Proceedings of the SID, vol. 27, no. 1, pp. 31–35, 1986.

18. K. Utsumi et al., Drop-on-demand ceramic ink-jet head using designed-space forming technology, *NEC Res. & Dev.*, no. 85, pp. 7–14, 1987.

19. J.C. Yang et al., A simple piezoelectric droplet generator, *Experiments in Fluids*, vol. 23, pp. 445–447, 1997.

20. A. Schober et al., Accurate high-speed liquid handling of very small biological samples, *BioTechniques*, vol. 15, no. 2, pp. 324–329, 1993.

21. T.P. Theriault, S.C. Winder, and R.C. Gamble, Application of ink-jet printing technology to the manufacture of molecular arrays, in *DNA Microarrays — A Practical Approach*, M. Schena, Ed., Oxford University Press, Oxford, 1999, chap. 6.

22. S.A. Elrod et al., Acoustic Lens arrays for ink printing, U.S. Patent 4,751,530, 1988.

23. B.B. Hadimioglu et al., Acoustic ink printing: printing by ultrasonic ink ejection, IS&T Eighth International Congress on Advances in Non-Impact Printing Technologies, H. Taub and E. Hanson, Eds., Society for Imaging Science and Technology, 1992, pp. 441–415.

24. B.B. Hadimioglu, C.F. Quate, and B.T. Khuri-Yakub, Liquid surface control with an applied pressure signal in acoustic ink printing, U.S. Patent 5,229,793, 1993.

25. I. Amemiya et al., Ink jet printing with focused Ultrasonic beams, Recent Progress in Ink Jet Technologies II, Eric Hansen, Ed., Society for Imaging Science and Technology, 1999, pp. 275–279.

26. S.F. Pond, *Inkjet Technology and Product Development Strategies*, Printhead design, Torrey Pines Research, Carlsbad, CA, 2000, pp. 83–151.

27. F.G. Tseng, Microdrop Generators, in *The MEMS Handbook,* Mohamed Gad-el-Hak, Ed., CRC Press LLC, Boca Raton, FL, 2002, pp. 30-1, 30-30.

28. E. Manini and A. Scardovi, The liquid spark inkjet technology, Proceedings of the Third International Congress on Advances in Non-Impact Printing Technologies, Joseph Gaynor, Ed., Society for Imaging Science and Technology, 1987, pp. 314–321.

29. M. Mutah, S. Kaieda, and K. Kamimura, An application of ink jet to the X-Y plotter, *J. Appl. Photographic Eng.*, vol. 5, no. 3, pp. 78–82, 1980.

30. D.H. Choi and F.C. Lee, Continuous gray-scale printing with the electrodynamic ink-jet principle, IS & T Eighth International Congress on Advances in Non-Impact Printing Technologies, H. Taub and E. Hanson, Eds., Society for Imaging Science and Technology, 1992, pp. 334–339.

31. G. Oda and M. Miura, The fundamental printing characteristics of a new electrostatic inkjet head, IS & T Eighth International Congress on Advances in Non-Impact Printing Technologies, H. Taub and E. Hanson, Eds., Society for Imaging Science and Technology, 1992, pp. 343–345.

32. D.H. Choi and F.C. Lee, Continuous-tone prints by the electrodynamic ink-jet method, The 9th International Congress on Advances in Non-Impact Printing Technologies'93, Y. Takahashi and M. Yokoyama, Eds., Society for Imaging Science and Technology, 1993, pp. 298–301.

33. J.L. Johnson, *Principles of Nonimpact Printing*, Palatino Press, Irvine, CA, 1998, pp. 328–331.

34. T.Murakami et al., High definition ink-jet printing: 10–20 um dots ejected from an injection needle, Recent Progress in Ink Jet Technologies II, Eric Hanson, Ed., Society for Imaging Science and Technology, 1999, pp. 263–274.

35. R. Carson and C.H. Hendricks, Natural pulsations in electrical spraying of liquids, *AIAA J.*, vol. 3, no. 6, pp. 1072–1075, 1965.

36. N.Kiyohiro and Y.Asako, Model for droplet ejection in the electro-rheological fluid Inkjet, IS & T Eighth International Congress on Advances in Non-Impact Printing Technologies, H. Taub and E. Hanson, Eds., Society for Imaging Science and Technology, 1992, pp. 340–342.

37. J. Fernandez de la Mora and A. Gomez, Remarks on the paper Generation of micron-sized droplets from the Taylor Cone, *J. Aerosol Sci.*, vol. 24, no. 5, pp. 691–695, 1993.

38. M. Cloupeau and B. Prunet-Foch, Electrohydrodynamic spraying functioning modes: a critical review, *J. Aerosol Sci.*, vol. 25, no. 6, pp. 1021–1036, 1994.

39. A. Jaworek and A. Krupa, Classification of the modes of EHD spraying, *J. Aerosol Sci.*, vol. 30, no. 7, pp. 873–893, 1999.

40. G. Percin, T. Lundgren, and B.T. Khuri-Yakub, Controlled ink-jet printing and deposition of organic polymers and solid particles, *Appl. Phys. Lett.*, vol. 73, no. 16, pp. 2375–2377, 1998.

41. J. Ducrée et al., TopSpot — A new method for the fabrication of microarrays, Proceedings MEMS 2000, Miyazaki, Japan, 2000.

42. N. Hey et al., TopSpot — The second generation of the new microarray method, 1st International Symposium Synthesis, Screening, Sequencing, Frankfurt, 2000.

43. B. de Heij et al., A tunable and highly-parallel picoliter-dispenser based on direct liquid displacement, Proceedings of IEEE-MEMS 2002 Las Vegas, 2002, pp. 706–709.

CHAPTER 3

Particle Kinetics of Ejected Microdrops

3.1 REYNOLDS NUMBER AND STOKES LAW

When designing a microdrop ejection system intended to target the drops accurately into predetermined locations, it is important to understand the factors governing the motion of microdrops in air. A microdrop is small enough that its associated Reynolds number puts the drop in a regime where the forces associated with viscous flow resistance to motion defined by Stokes Law is the dominant factor. The Reynolds number is a measure of the ratio of the dynamic pressure drag force to the Stokes viscous drag force on a given object moving through a fluid media such as air. This Stokes drag force is different from the more familiar dynamic drag force associated with macroscopic objects, such as aircraft, in that it is first order independent of air density and is proportional to the first power (not the square) of the speed of the object (Tables 3.1 and 3.2).

$$\text{Re} = vd/\nu \quad \text{Reynolds number}$$
$$(\text{Re} \propto \text{dynamic pressure force/viscous force}) \qquad (3.1)$$

where v = drop velocity relative to air
  d = drop diameter
  ν = Kinematic viscosity of air (0.151 cm^2/sec at standard temperature and pressure) = η/ρ
  η = viscosity of air (182.7 × 10^{-6} gram/cm-sec at standard temperature and pressure)
  ρ = density of air (0.001206 gram/cm^3)

The resistance caused by the air to the drop in its direction of motion caused by this viscous resistance is:

$$F_{Stokes} = 6\pi\eta rv \hspace{4cm} (3.2)$$

where v = drop velocity relative to air
   r = drop radius
   η = viscosity of air (182.7×10^{-6} gram/cm-sec at standard temperature and pressure)

The origin of this force is from the viscosity of the medium, in this case, air. Viscosity is roughly defined as the resistance of a medium to a change in its shape. The energy needed to orderly deform a viscous medium so that an object can pass through it (laminar flow) is the source of the Stokes resistance.

This is distinct from the more familiar conventional dynamic pressure forces associated with macroscopic objects such as aircraft (i.e., aerodynamic drag).

Table 3.1   Volume and Mass of Microdrops

Drop Diameter (Microns)	Drop Volume (Liters)	Drop Mass (Grams)	Drop Mass (Gev)
0.1	5.2×10^{-19}	5.2×10^{-16}	2.9×10^{8}
0.5	6.5×10^{-17}	6.5×10^{-14}	3.7×10^{10}
1.0	5.2×10^{-16}	5.2×10^{-13}	2.9×10^{11}
5.0	6.5×10^{-14}	6.5×10^{-11}	3.7×10^{13}
10.0	5.2×10^{-13}	5.2×10^{-10}	2.9×10^{14}
25.0	8.2×10^{-12}	8.2×10^{-9}	4.6×10^{15}
50.0	6.5×10^{-11}	6.5×10^{-8}	3.7×10^{16}
100.0	5.2×10^{-10}	5.2×10^{-7}	2.9×10^{17}

Table 3.2   Size Range of Microdrop Dimensioned Objects

Object	Size in Microns
Virus	0.1
Compacted DNA (5000 base pairs)	0.04–0.2
Inkjet pigments	0.1
IC manufacturing photolithography limit	0.2
Tobacco smoke	0.25
Bacteria	1
Open DNA strand (5000 base pairs)	1
Standard pigments	1–5
Red blood cell	8
SLAC automated Millikan drops	7–25
Typical animal cell	10
Flour dust	15–20
Inkjet printer drop	15–40
Pollen	15–70
Spray can mist	1–100
Human hair diameter	100

Table 3.3 Reynolds Numbers for Various Objects

Object	Approximate Reynolds Number
Airplane	10^8
Car at freeway speed	10^7
Thrown rock	10^5 Turbulent flow
Falling raindrop	10^3 Dynamic pressure dominates drag
Flying insect	10^1
100 μ water drop falling in air	1.8
50 μ water drop falling in air	0.3
20 μ water drop falling in air	10^{-2} Laminar flow
10 μ water drop falling in air	10^{-3} Stokes Law dominates drag
1 μ water drop falling in air	10^{-6}

$$F_{Drag} = \left(\frac{1}{2}\right)\rho C_d A v^2 \qquad (3.3)$$

where C_d = drag coefficient

ρ = air density

v = velocity relative to air

A = frontal area

In the case of dynamic pressure drag, the resistance to motion is caused by the random collisions of the particles making up the media with the object moving through it. The energy needed to accelerate the molecules of the media to the speed of the object moving through the fluid media is the origin of the resistance to motion.

There are significant differences in calculating the behavior of microdrops moving through air as opposed to macroscale objects such as aircraft and thrown rocks. For macroscale objects, the primary aerodynamic force is dynamic drag, which is proportional to the square of the object's speed and is proportional to the density of the air. For microdrops the principal aerodynamic force is Stokes drag, which is proportional to the first power of the drop speed and is independent of the air density.

Table 3.3 makes it clear that for objects of the sizes of micron-diameter fluid drops, the Stokes Law drag factor dominates over dynamic pressure drag in determining both the terminal velocity in air and the relaxation time constant.

There are secondary factors that need to be added to the simple form of the Stokes Law equation if highly accurate quantitative information is to be extracted from the motion of fluid microdrops.

Cunningham's correction factor (a.k.a. "Millikan resistance factor" or "slip") is a correction for the atmosphere not being a perfect continuum. The net result is that the resistance of the air is slightly less than that predicted by Stokes Law by a factor of $1/C_c$. The error is larger for smaller drops, about 16% for 1-micron-diameter drops and falls to a correction of less than 2% for 10-micron-diameter drops.

$$C_c = 1 + (2\lambda/d)(A + Qe^{(-bd/2\lambda)}) \quad \text{Cunningham correction factor} \qquad (3.4)$$

where A = 1.252

$Q = 0.399$
$b = 1.100$
$\lambda = 0.065$ microns
d = drop diameter in microns

3.1.1   Buoyancy Correction

This is a correction for the reduced effective weight of the drop due to the air that is displaced by the drop. The effect is minor for air at standard temperature and pressure since air has approximately 1/1000 the density of water.

3.1.2   Internal Drop Flow

As a fluid drop falls through the air, the fluid in the drop can internally circulate reducing surface resistance to air flow. Since the viscosity of most fluids is so much higher than the viscosity of air this correction to the terminal velocity of a falling drop is generally on the order of 0.5% or less.

3.1.3   Nonspherical Drops

Drops as they are falling can be deformed from perfect spheres, both by aerodynamic forces and by electric fields. A first order correction of this effect for drops with low Reynolds numbers is given by:

$$F = (3\pi\mu v d_n)k_n \qquad\qquad (3.5)$$

$$k_n = 1/3 + (2/3)(d_s/d_n) \qquad\qquad (3.5a)$$
non-spherical drop correction factor to Stokes Law

where d_s = diameter of a sphere with a surface area equal to that of the deformed
          drop
       d_n = diameter of a circle with the same projected frontal area in the direction
          of motion as the deformed drop

This is normally a negligible factor for drops below 100 microns in diameter.

3.2 TERMINAL VELOCITY

The terminal velocity in air of a fluid microdrop for drops is calculated by setting the gravitational force on the drop equal to that of the velocity dependent drag forces.
The terminal velocities for drops larger than 100 microns were taken from Reist.[1] The terminal velocities of drops smaller than 100 microns were calculated from Stokes Law including the Cunningham slip correction. Drops smaller than 0.1 microns in air have their movements dominated by Brownian motion. Studies on

Table 3.4   Terminal Velocity in Air of Fluid Drops (Density 1 gm/cm³)

Diameter (Microns)	Velocity (mm/sec)
0.1	9.0×10^{-4}
0.5	1.0×10^{-2}
1.0	3.5×10^{-2}
2.0	1.3×10^{-1}
5.0	7.7×10^{-1}
10.0	3.0×10^{0}
15.0	6.7×10^{0}
20.0	1.2×10^{1}
25.0	1.9×10^{1}
50.0	7.5×10^{1}
75.0	1.7×10^{2}
100.0	2.6×10^{2}
200.0	7.1×10^{2}
400.0	1.6×10^{3}
1000.0	4.0×10^{3}
5800.0	9.2×10^{3}

raindrops found that drops larger than 6 mm in diameter were broken up by aerodynamic forces into smaller diameter drops (Table 3.4).

Inkjet drop-on-demand devices typically eject their drops with an initial velocity of between one and ten meters per second. This is about three orders of magnitude faster than the final terminal velocities in air of the ejected droplets. For such applications understanding the nonequilibrium kinematics of the behavior of the ejected drops can often be highly important in designing drop-on-demand fluid deposition systems. How rapidly the drop decelerates is critical to determining what are usable drop ejector to target distances. The parameter one needs to know in order to characterize the motion of a fluid drop that has not come into equilibrium with externally applied forces is the relaxation time constant.

3.3 RELAXATION TIME CONSTANT

The relaxation time constant, τ, is a measure of how fast the motion of the drop comes to steady state after initial ejection, changes in local air speed or application of external forces induced by electric or magnetic fields. This is a very important number for designing fluid drop material deposition systems. The speed of drops at the ejection aperture can be in the meters per second range while the terminal velocity of the drops due to gravity may be in the millimeters per second range. This can result in large differences in the impact behavior of the drop depending upon how much the drop has slowed in air from its initial ejection velocity.

Also, a drop while moving just after ejection at high velocity is much less vulnerable to deflection by convection currents. Operation in totally convection free air may be impractical, particularly if the drop ejector is being continuously moved and repositioned by a robotically controlled motion stage. Placing the target material

Table 3.5   Relaxation Time Constant of Microdrops (Fluid Density = 1 gram/cm³) in Air

Diameter (Microns)	Relaxation Time (Seconds)	Horizontal Range (Ejected at 1 meter/sec) (mm)	Horizontal Range (Ejected at 10 meters/sec) (mm)
0.1	8.7×10^{-8}	0.000087	0.00087
0.5	1.0×10^{-6}	0.001	0.01
1.0	3.5×10^{-6}	0.0035	0.035
2.0	1.3×10^{-5}	0.013	0.13
5.0	7.8×10^{-5}	0.078	0.78
7.0	1.5×10^{-4}	0.15	1.5
10.0	3.1×10^{-4}	0.31	3.1
12.0	4.4×10^{-4}	0.44	4.4
15.0	6.9×10^{-4}	0.69	6.9
20.0	1.2×10^{-3}	1.2	12
25.0	1.9×10^{-3}	1.9	19
40.0	4.8×10^{-3}	4.8	48
50.0	7.6×10^{-3}	7.6	76
75.0	1.7×10^{-2}	17	170
100.0	3.0×10^{-2}	30	300

within this high velocity region depending upon operation requirements can preclude the need to operate in stagnant air.

The extent of the high velocity region of an ejected drop's trajectory can be obtained by equations 3.8 and 3.10 or more simply estimated by multiplying the relaxation time by the drop's initial ejected speed. This also gives the approximate distance that a drop will be projected if ejected horizontally. Since Stokes Law is implicitly assumed to provide the dominant breaking force on the drops, the horizontal range of the largest drops is likely to be less than what this simplified analysis predicts for large drops moving at high speeds (Table 3.5).

$$\tau = (1/18)(d^2/\eta)\rho \qquad \text{relaxation time constant} \qquad (3.6)$$

$$D_h = \tau v_{hi} \qquad \text{total horizontal distance traveled} \qquad (3.7)$$
$$\text{by a horizontally ejected drop}$$

where v_{hi} = horizontal initial ejected drop speed
  v_{vi} = vertical initial ejected drop speed
  d = drop diameter
  ρ = drop density
  η = viscosity of air (182.7×10^{-6} gram/cm-sec at standard temperature and pressure)
  g = gravitational acceleration

$$v_h(t) = v_{hi}e^{(-t/\tau)} \qquad \text{horizontal speed as a function} \qquad (3.8)$$
$$\text{of time}$$

$$D_h(t) = v_{hi}\tau(1 - e^{(-t/\tau)}) \qquad \text{horizontal position as a function} \qquad (3.9)$$
$$\text{of time}$$

$$v_v(t) = \tau g + (v_{vi} - \tau g)e^{(-t/\tau)} \qquad \text{vertical speed as a function of} \qquad (3.10)$$
$$\text{time}$$

$$D_v(t) = \tau gt + \tau(v_{vi} - \tau g)(1 - e^{(-t/\tau)}) \quad \text{vertical position as a function of} \qquad (3.11)$$
$$\text{time}$$

When designing a drop-on-demand material deposition system that must operate outside of a convection controlled chamber, it is advantageous to position the target such that the drop traverses the distance between the ejector and the target at a high enough velocity that local air currents do not significantly deflect the drop.

For very small drops this may be difficult to implement in practice. The deceleration is rapid, and the maximum stable monodisperse drop ejection speed tends to be slower the smaller the drop size. Ejection speeds cannot be arbitrarily increased by increasing the drive energy applied to the drop generator. Past a certain drive level, stable drop production ceases and the drop ejector starts to produce satellite drops and spray. Depending upon the fluid, the difference in drive energy between the threshold of drop ejection and the drive level that produces unstable ejection can be very small. This argues in general for the use of as large a drop as possible for any given application that necessitates accurate drop delivery in open air.

3.4 STREAMING

Drops moving through the air produce motion in the air that effects the motion of other nearby drops. This can be seen clearly when operating drop generators at high rates. One manifestation of interdrop coupling via the media (air) the drops are traversing is droplet streaming. Streaming is the tendency for droplets ejected at a high rate from a single source to collimate into a single well defined stream. When taking precision measurements of drop positions and velocities the motion of drops tens of diameters away can significantly effect the motion of the measured drop. This increases the velocity of the drop from what it would have in isolation. Streaming also collimates the drops tending to minimize deviations from the average trajectory.

3.5 IMPACT

The impact behavior of microdrops of less than 100 microns in diameter on a solid surface such as a microarray substrate is dominated by surface energy effects. For a drop to splash and break up on impact the kinetic energy must be sufficient to supply the surface energy for the new drops. For drops of water like fluids ejected from drop-on-demand devices with different diameters and impact speeds this condition applies for only relatively large drops traveling at high speed.

The surface energy of a drop is equal to the surface tension of the fluid times the surface area of the drop. The kinetic energy is equal to one half the mass times the velocity squared (Table 3.6).

Table 3.6   Surface Energy vs. Kinetic Energy of Ejected Microdrops

Drop Diameter (Microns)	Surface Energy (Surface Tension Water = 0.073 N/meters) (Joules)	Kinetic Energy	
		1 meter/sec (Joules)	10 meters/sec (Joules)
0.1	2.3×10^{-15}	2.6×10^{-19}	2.3×10^{-17}
0.5	5.7×10^{-14}	3.3×10^{-17}	3.3×10^{-15}
1.0	2.3×10^{-13}	2.6×10^{-16}	2.6×10^{-14}
5.0	5.7×10^{-12}	3.3×10^{-14}	3.3×10^{-12}
10.0	2.3×10^{-11}	2.6×10^{-13}	2.6×10^{-11}
20.0	9.2×10^{-11}	2.1×10^{-12}	2.1×10^{-10}
50.0	5.7×10^{-10}	3.3×10^{-11}	3.3×10^{-9}
100.0	2.3×10^{-9}	2.6×10^{-10}	2.6×10^{-8}
200.0	9.2×10^{-9}	2.1×10^{-9}	2.1×10^{-7}
500.0	5.7×10^{-8}	3.3×10^{-8}	3.3×10^{-6}

Since the surface energy of a single large drop is smaller than that of the same volume of fluid dispersed among smaller drops, the kinetic energy of the fluid drops with the exception of the largest drops at the highest speed is insufficient to break the drop up into smaller drops on impact. The microdrop on hitting a nonporous surface will remain a single cohered mass with a diameter determined by the surface tension of the fluid and the contact angle of the fluid with the substrate.

For aqueous fluids on metal and glass surfaces in most cases the diameter of the fluid spot will be from two to four times the drop diameter. For some fluid substrate combinations such as silicone oil and glass that have an essentially a zero contact angle, the fluid will spread out after impact forming an irregular boundary thin film. We have observed Newton's rings in deposits of silicone oil on glass.

3.6 BROWNIAN MOTION

Microdrops produced by drop-on-demand fluid jet devices are often in a size range where Brownian motion is a significant factor in the degree of accuracy with which one can track and predict the trajectory of drops ejected into air. For small drops the Brownian displacement over a few seconds of measurement time can be many times the diameter of the drop.

Where d = drop diameter
  η = viscosity of air (182.7×10^{-6} gram/cm-sec at standard temperature and pressure)
  T = temperature
  k = Boltzmann's constant 1.38×10^{-16} erg/Kelvin
  t = time in seconds

The mean distance that Brownian motion will deviate the trajectory of a particle from its predicted path over a time period, t, is given by:

Table 3.7   Brownian Motion Displacement
of Microdrops in Air

Drop Diameter (Microns)	Brownian Displacement	
	1 Second (Microns)	1 Minute (Microns)
0.1	2.2×10^1	1.7×10^2
0.5	9.7×10^0	7.5×10^1
1.0	6.9×10^0	5.3×10^1
5.0	3.1×10^0	2.4×10^1
10.0	2.2×10^0	1.7×10^1
20.0	1.5×10^0	1.2×10^1
50.0	9.7×10^{-1}	7.5×10^0
100.0	6.9×10^{-1}	5.3×10^0

$$\Delta x = (2Dt)^{1/2} \qquad (3.12)$$

where

$$D = kT/(\text{drag force}) \text{ is the diffusion coefficient} \qquad (3.12a)$$

For microdrops the drag force is given by Stokes Law so that the Brownian trajectory deviation is:

$$\Delta x = [(2kTt)/(3\pi\eta d)]^{1/2} \qquad (3.13)$$

As an example of the magnitude of Brownian motion on the trajectory of fluid drops in air is in Table 3.7.

3.7 MOTION IN ELECTRIC FIELDS

3.7.1   Electrostatic Deflection of Charged Microdrops

The force on a charged drop is proportional to the charge and the electric field at the drop.

Where v = drop velocity relative to air
   r = drop radius
   η = viscosity of air (182.7×10^{-6} gram/cm-sec at standard temperature and pressure)
   N_e = number of electron charges
   E = electric field (in volts/cm)

$$F_E = (1.6022 \times 10^{-12}) EN_e \quad \text{force on the charged drop in dynes} \qquad (3.14)$$

**Table 3.8   Terminal Velocity of Charged Microdrops in an
Electric Field of 1000 Volts/cm**

Drop Diameter (Microns)	Δv/Charge (cm/sec)	(Δv Charge) × (N_e max from Rayleigh Limit) (cm/sec)
0.1	9.3×10^{-2}	1.3×10^2
0.2	4.7×10^{-2}	1.8×10^2
0.5	1.9×10^{-2}	2.9×10^2
1.0	9.3×10^{-3}	4.1×10^2
2.0	4.7×10^{-3}	5.9×10^2
5.0	1.9×10^{-3}	9.3×10^2
10.0	9.3×10^{-4}	1.3×10^3
20.0	4.7×10^{-4}	1.9×10^3
50.0	1.9×10^{-4}	2.9×10^3
100.0	9.3×10^{-5}	4.1×10^3

When operating in air, the resistance to the motion is given for low Reynolds numbers by Stokes Law.

$$F_{Stokes} = 6\pi\eta rv \qquad\qquad (3.15)$$

The alteration of the trajectory of a charged microdrop can be considerable even for modest applied electric fields. Table 3.8 summarizes the effect of an electric field of 1000 volts/cm on microdrops moving in air under standard temperature and pressure.

The extrapolated maximum velocity changes obtained by multiplying the ΔV per unit charge by the maximum charge that one can place on the drop without exploding it slightly overpredicts the maximum speeds since the drops in many cases are moving fast enough that Stokes Law no longer accurately predicts the drag force on the fluid microdrop. The order of magnitude of the numbers does indicate that control over the motion of charged drops with externally applied electric fields is eminently practical. Control over the motion of charged microdrops using modulated electric fields is used in commercially successful devices such as cell sorters and inkjet printers.

One also notes from these numbers that the reverse process, extracting the value of the electric charge on fluid microdrops from their motion in an electric field, is also very practical. In fact, using modern machine vision systems, experiments being conducted at SLAC to detect fractionally charged fundamental particles are able to resolve the charge on 10-micron diameter microdrops to 1/50 of an electron charge.[2]

REFERENCES

1. P.C. Reist, *Aerosol Sci. & Tech.*, 2nd ed., McGraw Hill, New York, 1993.
2. V. Halyo et al., Search for free fractional electric charge elementary particles using an automated Millikan oil drop technique, *Phys. Rev. Lett.*, vol. 84, no. 12, pp. 2576–2579, 2000.

CHAPTER 4

Electric Charging of Microdrops

Fluid microdrops are very easily given large electric charges that can then be used to control their positions and velocities. Even if one does not intend to use electric fields to control the motion of the ejected drops, drop charging is not a phenomenon that can be safely ignored. Drops that are produced with large amounts of charge can explode upon slight evaporation or be drawn back into the ejection aperture surface by electrostatic attraction after breaking off from the fluid jet. Interdrop electrostatic forces can affect the stability of ejected droplet streams and droplet arrays.

4.1 RAYLEIGH LIMIT TO CHARGING MICRODROPS

The limiting charge that can be applied to microdrops is the magnitude at which either spontaneous electron or ion emission occurs, or until the internal repulsion from the like charges exceed the surface tension. At this point, the drop explodes. The amount of acceleration inducible by electrostatic forces on a microdrop is limited by the Rayleigh limit for the amount of charge that will explode a drop and the magnitude of the electric field that will initiate spark or corona break down at the deflection electrodes.

Where d = diameter of fluid drop

N_e = number of electron charges

γ = fluid surface tension

e = electron charge

N_R = Rayleigh limit on the number of charges on a fluid drop

Table 4.1   Rayleigh Limit to Charging Microdrops
(Water Drops $\gamma = 72.7$ dyne/cm)

Drop Diameter (Microns)	Number of Charges (Rayleigh Limit)
0.1	1.4×10^3
0.5	1.6×10^4
1.0	4.5×10^4
2.0	1.3×10^5
5.0	5.0×10^5
10.0	1.4×10^6
20.0	4.0×10^6
50.0	1.6×10^7
100.0	4.5×10^7

$$N_R = (1/e) \, (2 \, \pi \gamma)^{1/2} \, (d)^{3/2} \tag{4.1}$$

Equation 4.1, known as the Rayleigh limit, gives the number of charges that a fluid drop can have and not disintegrate due to the electrical repulsion overcoming surface tension (Table 4.1). There is an additional limit on the amount of charge that a fluid drop can have, and this is the point where the local electric field is high enough that electrons or positively charged ions are spontaneously emitted from the drop. This limit cannot be reached before the drop explodes due to the Rayleigh limit except for drops less than 0.01 microns in diameter. Since the charge limit is proportional to the 3/2 power of the diameter, it is implied that once the Rayleigh limit is reached and the drop starts breaking up, it will continue to disintegrate until it reaches the limit of spontaneous electron emission, which, for negatively charged drops, is in the range of 0.1 to 0.01 microns in diameter depending upon fluid surface tension. Positively charged drops will disintegrate down to the molecular dimensions due to the higher electric fields required for positive ion emission.

4.2 DROP CHARGING MECHANISMS

The operating fluid that is jetted out of the ejection nozzle to form into a free flying microdrop can be thought of as leaky dielectric, in which the distribution of charges in the form of mobile ions can be readily shaped by stray and deliberately applied electric fields as well as material contact potentials. As a result, fluids in general are ejected from drop-on-demand devices with difficult to predict magnitudes of electric charges. There are numerous mechanisms operating together that act to effect the electrical charges on fluid microdrops.[1-3] Some of the major effects on the charges of ejected microdrops are from:

- Triboelectric charging
- Fluid jet polarization
- Electrical double layer charging
- Spray charging
- Statistical fluctuations

- Absorption of gaseous ions
- Photoelectric emission

4.2.1   Triboelectric Charging

Triboelectric charging is also known as contact charging. The production of a charged object comes about by the contacting and separation of materials with dissimilar affinity to electrons. As the fluid jets through the ejection aperture hole and the air, it can acquire an electric charge. The application of contact angle changing thin films onto the ejection aperture surface can have a large effect of the charges of the ejected fluid drops. However, the effect on the charge of the drops is not constant over time. This is a complex system to attempt to predict charging effects since small amounts of contamination, such as the debris built up from fluid debris, can change the electron affinity of the surface. On an ejection aperture surface having an insulating surface layer, a charge can build up that can counteract the original tendency to charge the ejected drops.

As an example, one drop ejector that SLAC researchers constructed had a Teflon® (E.I du Pont de Nemours and Company, Wilmington, Delaware) coated silicon micromachined ejection aperture that, when first loaded with distilled water, ejected the drops with such a high level of charge that the drops after breaking off from the fluid jet and decelerating to terminal velocity were pulled back to the ejection aperture surface. The drops gradually became less charged over a period of a couple of hours. Attempting to manipulate the charge of ejected drops by the use of thin films over the ejection aperture surface is unreliable in practice due to the large effects of surface contamination from the fluid itself. Eventually, if the surface of the ejection aperture becomes contaminated with a thin layer of the fluid, the surface ceases to become electrically dissimilar.

4.2.2   Fluid Jet Polarization

A conductive fluid projected from the ejection aperture hole into free space in which there is an electric field will have the ions in the fluid jet experience a force that will polarize the fluid jet cylinder. When the end of the jet breaks off to form a free fluid drop, that end will then be charged due to the ion imbalance over the length of the fluid cylinder. This electric field can be from stray electrostatics or it can be deliberately induced by an external charge control electrode in order to modulate the charges of the ejected drops. The electric field needed to induce a given charge on a drop is a function of the conductivity of the fluid. Aqueous fluids with a high ion concentration can be charged to near the Rayleigh limit with only a few hundreds of volt per cm field. The charge induced is a roughly linear function of the applied electric field. The spread in the value of the charge is quite small as a percentage of the total induced charge. This method is used in commercial cell sorters to differentially charge their fluid drops. The time constant for changes in the induced drop charges in the case of highly conductive aqueous fluids, is also very short on the order of microseconds or less. In contrast, in insulating fluids such as in silicone oil, altering the charges of the ejected drops required thousands of volts per cm and the duration for equilibration of charge after a change in the induction voltage was hours.

4.2.3 Electrical Double Layer Charging

Electrical double layer is the term for the formation of a high concentration of charge along a fluid-to-solid interface by the attraction of ions or polar molecules to a surface with a different affinity for electrons. This produces an ion concentration gradient that can cause an imbalance in the positive and negative ions in the ejected drops. This is the source of the streaming potential that is produced when a fluid containing mobile ions is forced through a capillary or porous membrane.

4.2.4 Spray Charging

Spray charging electrically charges microdrops in a manner similar to electrical double layer charging. In this case, the electrical double layer is at the fluid air interface. The formation of a fluid drop by the breaking off of a portion of a fluid air interface where charge has been segregated will result in a drop with a net electrical charge.

4.2.5 Statistical Fluctuations

For most applications that require thousands of electron charges in order to produce motion with a minimum applied electric field, the statistical spread caused by thermal noise and quantization is not significant. These effects can become important for some scientific applications where near neutrality to within ten electron charges or less are needed. Since charges are quantized, any random volume of fluid can have an imbalance at any given point in time in the amount of positive and negative charges. The amount of random charge imbalance that can be sustained against the resultant electric field is on the order of the temperature in Kelvin multiplied by Boltzmann's constant.

4.2.6 Absorption of Gaseous Ions

After the drop is ejected, its charge can still be altered. One mechanism is the introduction of ionized gas molecules in the air through which the drops must pass. This is the basis for the operation of most commercial aerosol neutralization units. The two most common sources of ions for this purpose are corona discharge and radioactive sources. The exact neutralization of the microdrops to a mean charge of zero can be complicated in some circumstances by the different mobilities and lifetimes of the positive and negative atmospheric ions.

4.2.7 Photoelectric Emission

In principle, it is possible for photons of a sufficiently short wavelength to eject electrons from the surface of a fluid drop in order to alter its charge. In practice, this is difficult to implement in isolation with other charge changing processes because the wavelengths required for most fluids are in the sub-300 nanometer ultraviolet range, which can ionize air, produce ozone, as well as eject electrons from the walls of the containment vessel.

4.3 PRACTICAL DROP CHARGE CONTROL METHODS

4.3.1   Contact Surface Modification

One method that can control the charge on ejected microdrops is to apply thin film surface coatings to the ejection aperture surface to utilize triboelectric charging and double layer charging. However, this is not practical because contact angle optimization and anti-abrasion characteristics of the nozzle surface take design precedence. Also, as the surface becomes contaminated with use, the charges of the droplets will change by large amounts over that of a fresh, thin film coating. The fact that the triboelectric and double layer charging will change over time affects the uniformity of induction charging since this takes place in addition to fluid jet polarization charging.

4.3.2   Fluid Jet Polarization

In practice, the techniques one uses to obtain drops of a given charge range will depend upon the magnitude of the charge that one requires. For obtaining charges in the thousands to hundreds of thousands of electron charge range, the technique used universally in commercial printing and droplet sorting devices is charging by fluid jet polarization. In fluid jet polarization, the region into which the drop-forming fluid jet is ejected, has a set of electrodes that applies a controllable electric field along the length of the fluid jet. This causes charge separation along the length of the jet resulting in the tip of the fluid jet, when breaking off, having a net nonzero electric charge. This situation is roughly analogous to a capacitor charging scenario in which the length of the fluid jet is the resistor and the tip of the jet, which forms the drop, is the capacitor.[4] Not all fluids can be effectively charge controlled in this matter. One way of estimating whether a fluid can be effectively charged by an externally applied electric field is to define a characteristic time for charge changes in the liquid and compare it to the time scale over which fluid jet ejection and droplet breakoff takes place.

Where ε_r = dielectric constant
ε_0 = permittivity constant of free space
n = ion density
μ = ion mobility
d = drop diameter
T = temperature
e = elementary charge

$$t_c = (\varepsilon_r \varepsilon_0 / ne\mu) \text{ characteristic time for charge changes} \qquad (4.2)$$

For water in which the hydronium ion is the principle charge carrier, the characteristic time is on the order of a microsecond. For mineral oils in which the principal charge carriers are charged surfactant micelles, the characteristic time is on the order of a millisecond. Since the drop ejection process takes on the order of a few tens of microseconds, one would predict that charge control using fluid jet polarization would be very effective on aqueous fluid but ineffective on insulating oils. This is in fact what has been observed in experiments. This does not mean that the charge of the drops of ejected oils cannot be influenced by an applied external electric field but that the electric fields must be higher for a given induced charge. There is also a time delay before the full influence of an externally applied electric field on the drop charge is seen. Drops will become more and more charged until they stabilize at a relatively fixed level of charge over a period of minutes to hours. This result has been observed with Dow Corning silicone oil (Dow Corning Corporation, P.O. Box 994, Midland, MI) and Octoil®-S diffusion pump oil (CVC Products, Inc., Rochester, NY), but the mechanisms are fully understood.

Similarly, an estimate on the actual magnitude of the charge induced on a high conductivity fluid during the ejection process can be given[5] by a calculation based on the potential of the drop at break off within a defined electric field:

Where V = induction electrode potential
        L = orifice droplet distance at break off from the fluid jet
        H = induction electrode to ejection aperture distance
        e = elementary charge
        k = Boltzmann's constant
        ε_r = dielectric constant
        ε_0 = permittivity constant of free space
        d = drop diameter
        σ = fluid conductivity
        α = electric field geometry factor

$$n_Q \simeq -(2\pi\varepsilon_0\, d\, \alpha\, V\, L)/eH \text{ number of electron charges induced} \qquad (4.3)$$

Example 4.1   Induced Charge on Microdrops

As an example for a realistic system with the parameters:

Where L = jet length at break off equal to five times the drop diameter
        H = 0.2 cm induction electrode to ejection aperture gap
        $\alpha = 1$
50 volt per mm electric field along fluid jet

The electrical conductivity is assumed to be high enough that the charge equilibration time is short compared with the fluid jet ejection and drop formation time (~ 10 microseconds) (Table 4.2).

Table 4.2 Induced Charge on an Ejected Microdrop as a Function of Drop Diameter

Drop Diameter (Microns)	Induced Charge (Electron Charges)
1.0	9×10^1
5.0	2×10^3
10.0	9×10^3
20.0	3×10^4
50.0	2×10^5
100.0	9×10^5

Since the maximum induction field that will not arc or initiate corona in air is on the order of 1000 volt/mm, it is clear that electrical induction drop charging systems can be quite effective for microdrops made from conductive liquids. This equation also indicates that the charge induced on the drop is influenced by variations in the actuator drive since this will vary the jet behavior, in particular the speed of the jet and its length at droplet break off. This also makes charging dependent upon fluid rheological properties since variations in viscosity and surface tension strongly influence the length of the jet at time of droplet break off. The estimate in the example of a jet with a length of five times the drop diameter at time of break off is typical of aqueous low viscosity liquids. However, there are jets with length-to-diameter ratios on the order of hundreds with certain polymer solutions. Conversely for high-surface-tension, low-viscosity fluids, such as pure water, drop ejection has been observed with no discernable connecting ligament at drop break off.

Example 4.2 Charging Time Constants for Common Fluids

For fluids with poor electrical conductivity charge, equilibrium may not have time to establish before drop break off takes place. One way to gain some quantitative insight into whether a fluid and drop ejection system is suitable for a charge control application is to model the system as an RC circuit in which the tip of the fluid jet is the capacitor and the length of the jet is a resistor. Using this model for drop charging, the RC time constant for this system is

$$t \simeq 8 \ (L/D) \ (\varepsilon_r \varepsilon_0 / \sigma) \tag{4.4}$$

Using the previous assumption of an L/D of 5 at drop break off:

Table 4.3 Electrical Conductivity and Charging Time Constants for Various Fluids

Fluid	Conductivity (S/meter)	ε_r	Time Constant (Seconds)
Distilled water	1×10^{-5}	80	3×10^{-3}
Tap water	3×10^{-2}	80	9×10^{-7}
Ethylene glycol	2×10^{-4}	41	7×10^{-5}
Sea water	5×10^{0}	80	6×10^{-9}
Silicone oil	10^{-15}	2	7×10^{5}

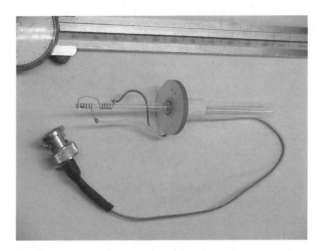

Figure 4.1   Drop ejector with charge control electrodes. The drop ejector shown has electrodes mounted along the end of its glass reservoir tube to allow for controlling the mean charge of its ejected drops using droplet jet polarization charging.

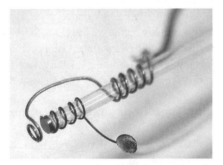

Figure 4.2   Details of the charge control electrodes. An electric field is induced along the ejected fluid jet by placing a potential difference between the ejection aperture, which is attached to the electrical circuit with a spring tensioned contact wire and a wire electrode ring through which the drops are ejected.

These numbers probably are uncertain to within an order of magnitude but they do confirm that for aqueous ionic fluid the charging time constants are short compared with the 10^{-5}-second time scale for droplet jet formation. These numbers also confirm observations from experiments that insulating oils are very poorly charge controlled by electric field induction (Table 4.3).

Modification of a drop-on-demand ejector for fluid jet polarization charge control can be implemented by attaching an electrode to the ejection aperture and placing an electrode ring in the ejection direction of the fluid jet (Figures 4.1 and 4.2).

4.3.3   Charge Control Using Ionized Air

The charge of a microdrop after ejection can be controlled by passing the drop through ionized air.[6] This is done primarily for neutralization of highly charged drops, but it also can be used for drop charging by manipulation of the local ion concentration by the use of externally applied electric fields.

Air can have ions introduced by ultraviolet light, radioactive sources, or corona discharge. We found that in practice properly shielded radioactive sources were environmentally more benign than ionization sources based on UV light and corona discharge systems due to the difficulty involved in dealing with the ozone produced by short wave UV and electric discharges.

The principle behind charge neutralization by passage of the drop through ionized air is that the ions with a charge opposite to those of the drop will be attracted, and the ones with a like charge will be repelled. The drop will eventually end up near neutral by this process with a final charge spread determined by the temperature and drop radius. A deviation from neutrality is possible because the random thermal energy of ions at room temperature is sufficient to break through the electrostatic repulsion of the drop.

The mobilities of atmospheric ions are in the range of 1 to 2 cm^2/volt-sec. Positive ions have a mobility that cluster around 1 cm^2/volt-sec while negative ions range between 1.5 to 2 cm^2/volt-sec. This velocity versus electric field range makes it very practical to use externally applied electric fields to alter the concentrations of positive and negative ions to control the average charges of drops passing through a certain region of ionized air.

Air ionization for droplet charge changing can be accomplished by radioactive sources, corona discharge, or ultraviolet light. Most industrial devices for aerosol neutralization use alpha- (helium nucleus) or beta- (electron) emitting radioactive sources such as Krypton-85, Polonium-210, Americium 241, and Strontium 90. The strength of the sources is in the range of hundreds of microcuries to a few tens of millicuries. Radioactive source strength varies due to differing requirements for volume ionization and neutralization time.

Where k = Boltzmann's constant
ε_r = dielectric constant
ε_0 = permittivity constant of free space
n+ = positive ion density
n− = negative ion density
μ+ = positive ion mobility
μ − = negative ion mobility
d = drop diameter
T = temperature in Kelvin

The standard deviation in the charge spread in the neutralized drops is:

$$\sigma = [(2 \pi \varepsilon_0 \, d \, k \, T)/e^2]^{1/2} \text{ electron charges} \qquad (4.5)$$

If it is desired to give the drops a net average charge, then one can alter the relative concentrations of the positive and negative ions, introduce an electric field or alter the gas composition so that the positive and negative ions have strongly differing mobilities.

The ion concentration, however, affects the rate at which equilibrium is approached, which occurs on a time scale comparable with:

$$t_{eq} = \varepsilon_0/(ne\mu) \text{ seconds} \tag{4.6}$$

where n is the ion concentration and μ is the ion mobility.

If there are significant concentration and mobility between the positive and negative ions, the average charge on the drops will deviate from neutrality from which the mean charge can be calculated:

$$n_q = \sigma^2 \ln[(n-\mu-)/(n+\mu+)] \tag{4.7}$$

where

$$\sigma = [(2 \pi \varepsilon_0 d k T)/e^2]^{1/2} \tag{4.7a}$$

Some design rules of thumb for ionized air neutralization/charging systems using radioactive sources:

- With beta-emitting sources, such as Sr-90 and Kr-85, the energy absorption coefficient is on the order of 0.04 cm^{-1} and about 200 ion pairs are produced per cm per decay.
- With alpha-emitting sources, such as Am-241 and Po-210, the range of the emitted particles is only on the order of a few cm but each emitted alpha particle generates about 4000 ion pairs per cm.

Both the beta and alpha particles are easy to shield but their passage through matter generates gamma radiation, which requires far more material to stop than alphas and betas.

Recombination of the ions with each other, induced air flow through the neutralizer, and absorption on the walls of the neutralizer containment vessels prevent an infinite build up of ion density. The ion density that a given source will produce is strongly dependent upon the specific physical design of the neutralizer. In order to get a feel for the magnitude on the ion density produced for various real world aerosol neutralizers, in 1973, Cooper and Reist[7] detailed that a neutralizer using a 5-cm long cylindrical Kr-85 beta source with a strength of 10 millicuries in an environment that exchanges out its volume of air each second, produced ion pair densities on the order of 10^7 pairs/cm.^{3} The theoretical neutralization time for a droplet in this environment is about 0.01 seconds.

Due to mobilities of the various ions generated by the radioactive source, and stray electric fields which can affect local ion concentrations, it has been found that a passively designed gas ionization charge neutralizer will only reduce the drop charges down to some mean value that may be a few tens of charges to a low hundred of an electron charge from a mean charge of zero. For many applications, the reduction of charges in the tens of thousands to these values may constitute effective neutralization. If neutralization to a mean value of exactly zero charge is desired,

Table 4.4   Minimum Variation in Induced
Microdrop Charge

Drop Diameter (Microns)	Minimum Charge Spread (Electron Charges)
1.0	3
5.0	7
10.0	9
20.0	13
50.0	21
100.0	30

researchers found that it was necessary to place electric field bias plates inside the ionization region to control the local ratios of positive to negative ions. Internally controlled bias fields on the order of a few volts per millimeter were found to be necessary to tweak the mean values of the drop charges to exact neutrality.

References 7–9 contain detailed designs for aerosol particle charge neutralizers.

4.4 VARIATIONS IN THE INDUCED CHARGE

There is a minimum spread in the induced charges no matter how reproducible the drop ejecting process and how precisely controlled the applied electric field. This is due to thermal noise which can supply the energy to displace charges above that of the electric field induced charge displacements (Table 4.4).

The standard deviation of the charge around the mean value is:

$$\sigma = [(2 \pi \varepsilon_0 d k T)/e^2]^{1/2} \qquad (4.8)$$

This limit also applies when gaseous ions are used to neutralize charged microdrops. It is possible though to beat this limit under very restricted conditions. One is to use an ion free fluid. In one experiment 10-micron diameter drops were formed using a sample of particularly pure Dow Corning silicone oil in which over 90% of the drops were ejected electrically neutral with most of the charged drops having only a charge of plus or minus one electron charge. This was, however, difficult to maintain and replicate on demand. Very small amounts of contaminants would render the ejected drops charged with a spread approximating the theoretical thermal limit. The other condition under which this limit can be cheated is to use controlled neutralization of individual drops. Tests by various research groups who used levitated solid microspheres to search for fractional electric charge utilized ultraviolet light and radioactive sources to alter the charges of their test objects to the near-neutral range, where their instruments were most sensitive. This charge control technique is, of course, dependent upon the ability to isolate and measure the charges of the individual drops to the precision that one wishes to set their charges. These charge spread limits are important in certain precision charge measurement experiments but are of little concern for drop trajectory control application in which the induced charges are typically on the order of thousands to tens of thousands of electron charges.

REFERENCES

1. P.C. Reist, *Aerosol Sci. & Tech.*, 2nd ed., McGraw Hill, New York, 1993.
2. Y. Gu and D. Li, Measurement of the electric charge and surface potential on small aqueous drops in the air by applying the Millikan method, *Colloids & Surfaces A: Physiochemical & Eng. Aspects*, vol. 37, pp. 205–215, 1998.
3. M. Polat, H. Polat, and S. Chander, Electrostatic charge on spray droplets of aqueous surfactant solutions, *J. Aerosol Sci.*, vol. 31, no. 5, pp. 551–562, 2000.
4. A. Atten and S. Oliveri, Charging of drops formed by circular jet breakup, J. *Electrostatics*, vol. 29, pp. 73–91, 1992.
5. G. Reischl, W. John, and W. Devor, Uniform electrical charging of monodisperse aerosols, *J. Aerosol Sci.*, vol. 8, pp. 55–65, 1977.
6. B.Y.H. Liu and D.Y.H. Pui, Equilibrium bipolar charge distribution of aerosols, *J. Colloid & Interface Sci.*, vol. 49, no. 2, pp. 305–312, 1974.
7. D.W. Cooper and P.C. Reist, Neutralizing charged aerosols with radioactive sources, *J. Colloid & Interface Sci.*, vol. 45, no. 1, pp. 17–26, 1973.
8. K.K. Whitby, Generator for producing high concentrations of small ions, *Rev. Sci. Instr.*, vol. 32, no. 12, pp. 1351–1355, 1961.
9. S.V. Teague, H.C. Yeh, and G.J. Newton, Fabrication and use of Krypton-85 aerosol discharge devices, *Health Phys.*, vol. 35, pp. 392–395, 1978.

CHAPTER 5

Engineering Requirements for Reliable Microdrop Generation

The process of drop ejection is not as simple as taking a fluid chamber with a small hole and pressurizing it enough for fluid to start emerging from the ejection nozzle hole. Attempting to eject a drop in this manner would not result in the ejection of a microdrop but rather in fluid accumulating at the aperture hole, spreading out on the surface, and increasing in area and volume until a macroscopic drop similar in size to a water faucet drop breaks off by gravity. A higher, steady pressure would result in a continuous stream of fluid being pushed out, ultimately breaking up into randomly sized droplets.

Drop-on-demand production of discrete microdrops requires very special conditions for monodisperse drops to form. The microdrop formation process, by which drop-on-demand devices, such as inkjet printers, generate their drops, require short-microsecond-duration, high-speed fluid jets. These short-duration fluid jets are produced in response to positive and negative pressure waves occurring within tens of microseconds of each other in the fluid internal to the ejection nozzle. The fluid near the ejection nozzle experiences accelerations on the order of a 100,000 Gs with the fluid jet speeds being of one-to-tens-of-meters-per-second (Figures 5.1 and 5.2).

5.1 CONDITIONS NEEDED FOR MONODISPERSE DROP-ON-DEMAND OPERATION

The conditions under which a drop-on-demand ejector fluid jet will form monodisperse microdrop are part of a very small subset of all possible operating conditions. The operating conditions must be tuned, usually empirically, by trial and error for proper drop-on-demand operation to occur.

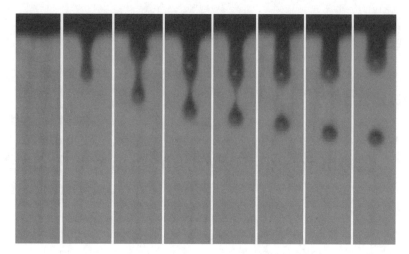

Figure 5.1   Time lapse images of drop-on-demand microdrop generation. These stro-
boscopically imaged pictures show the process of drop ejection at 20-microsec-
ond intervals for the ejection from a piezoelectrically actuated drop generator
ejecting a 50 micron diameter drop of a mixture of propylene glycol and water.
A fluid jet is pushed out at a high enough speed to form a cylindrical column. A
combination of Rayleigh Taylor instabilities and deliberately tuned standing waves
cause the fluid column to destabilize and have an end node condense into a
separate drop. This end node breaks off as the fluid column withdraws back into
the ejector as a result of a negative pressure wave on the interior of the drop
generator fluid chamber.

Assuming that the operating fluid has rheological characteristics that allow
monodisperse drops to be produced, the operating parameters for reliable microdrop
ejection in drop-on-demand devices that must be optimized are:

- Drive pulse amplitude
- Drive pulse shape
- Internal pressure level
- Drop ejection rate
- Fluid fill level

5.2 DRIVE PULSE AMPLITUDE

At amplitudes below the threshold of drop ejection, a fluid jet consisting of a
cylinder of liquid is ejected from the aperture and drawn back on the negative pressure
cycle of the pressure excitation. In extreme cases, this fluid jet can be in the milli-
meter-length range, depending upon the properties of the fluid. If the amplitude of
the ejection pulse is too small, the fluid jet will be completely withdrawn into the
dropper before the jet has destabilized enough to form discrete, free-flying drops.

If one has designed the ejector, drive signal, and the operating fluid properly,
there will be a drive voltage amplitude range that will eject a single drop with each
pulse of uniform size and identical velocities. Generally within this operating win-

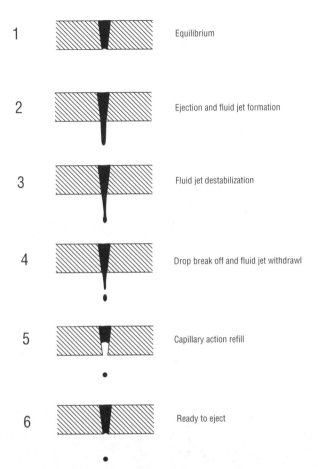

1   Equilibrium

2   Ejection and fluid jet formation

3   Fluid jet destabilization

4   Drop break off and fluid jet withdrawl

5   Capillary action refill

6   Ready to eject

Figure 5.2   Microdrop generation process in a drop-on-demand drop ejector. Microdrop ejection from drop-on-demand devices is a high speed complex process that starts with a fluid jet being shot out of the ejection aperture, the jet destabilizing via the Rayleigh Taylor instability and forming nodes along the jet, a negative pressure wave withdrawing the jet back into the drop ejector and the leaving of a single drop of the desired size ejected traveling in the desired direction.

dow, increasing the drive amplitude both increases the size of the drop and the ejection velocity (Figure 5.3).

The ejection speed of the drop as it breaks off from the fluid jet will usually be in the range of 1- to 10-meters-per-second. For small drops ejected into air, there will be a rapid deceleration following this initial high-speed ejection phase.

If the amplitude of the drop is too large, multiple drops will form producing satellite drops (Figure 5.4). However, this may be a stable, repeatable process. Satellite drop formation in some applications may not be fatal due to its repeatability and the common direction of travel of the primary drop and the satellites.

An even larger overdrive amplitude will throw a chaotic spray of randomly sized drops with random directions of travel (Figure 5.5).

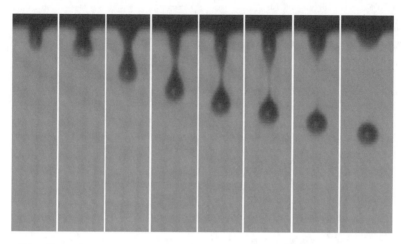

Figure 5.3   Formation of a microdrop from a fluid jet. At the proper drive amplitude the fluid jet will be ejected, destabilize and form a single free drop as the fluid jet retracts.

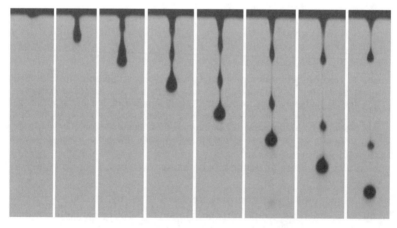

Figure 5.4   Formation of a microdrop with accompanying satellite drops. As the drive amplitude is increased past the threshold where the ejection of a single free drop takes place, the drop will be ejected with an increasingly higher initial speed. Past a certain jet velocity, multiple nodes will form on the jet and condense into additional discrete drops ejected simultaneously with the main microdrop. The generation of these satellite drops can be a stable, repeatable process. The location and size of these satellite-forming nodes can be altered by varying the excitation waveform, though prediction a priori of the fluid jet shape from the excitation waveform is, in general, not possible.

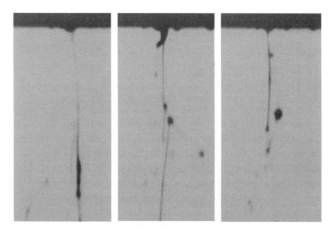

Figure 5.5   Ejection of random fluid sprays when overdriving a drop ejector. At a high enough drive amplitude, a chaotic spray of fluid drops, rather than a well-formed fluid jet, will be ejected from the fluid nozzle. The drop-spray pattern that is formed from this type of overdrive condition is different from drive pulse to drive pulse even with identical excitation wave forms. This is not a good way to operate a drop ejector, even momentarily. Overdrive conditions can cause air ingestion affecting reliability, even after the drive amplitude is reduced to more optimal levels. Also, chaotically produced spray can be drawn back to the ejection nozzle surface by electrostatic attraction depositing debris over and near the ejection nozzle hole.

Each ejection pulse at this level of overdrive will produce a different random mix of drops and ejection directions. This type of ejection pattern is usable however for applications such as aerosol generation, micromixing, and dispersion of reagents where the area of desired deposition is much larger that the spray radius.

One caution to be noted is that extreme overdrive conditions, depending upon the geometry of the fluid nozzle, may cause air ingestion into the interior of the drop generator and thereby a steady degradation of operation. The ejected spray can also contain small, highly electrically charged droplets that may redeposit onto the ejection aperture surface, forming a layer of fluid and debris that may impair drop ejector operation. Operating a microdrop generator in extreme overdrive conditions, even momentarily, may make it unable to subsequently produce directionally stable monodisperse drops without a cleaning and air purging operation.

5.3 DRIVE PULSE SHAPE

The pulse shape of the waveform used to excite a piezoelectrically actuated microdrop ejector has wide-ranging effects on the drop ejection process. The tuning of a particular drop ejector design for a given fluid by optimizing the drive waveform is a critical part of setting up a drop ejector system. This is an empirical observation that has been reported by many sources[1-3] with attempts also to predictively model the waveshape tuning process.[4] The fundamental reasons may be the exciting of different vibrational modes in the ejected jet and the mechanical structure of the drop ejector. These are difficult to model in a way that can predict, for fluids with a given rheological set of characteristics, what pulse width will be optimal for stable drop ejection. In practice settings, stable, monodisperse drops are found by trial and error and are different for different fluids and even for different fluid-fill levels within the same drop ejector. In general, these are some observations useful for tuning drop generators:

- Pulse widths between 0.5 microseconds and 10 microseconds were found to be the most often used range for large reservoir piezoelectric disc driven drop ejectors. Short reservoir tube cylindrical piezoelectric drive drop ejectors functioned best with drive pulses in the 5 microsecond to 20 microsecond range.
- Shorter pulse widths tended to produce smaller drops. This is a general trend only and is far from a one-to-one or even a monotonic relationship.
- Shorter pulse widths tend to require higher amplitudes to eject drops. Again, this is not absolutely true for all cases. In fact, local amplitude minimums will be found for drop ejection if one systematically scans through all pulse widths to find the threshold for drop ejection. The differences in amplitude for drop ejection can be as high as three to one over the pulse width range between 1 and 10 microns. The minima often occur not at the ends of the range but at islands with widths that can be as narrow as a few tenths of a microsecond.
- Some pulse widths have no amplitude window at which stable monodisperse drops are produced. The minimum amplitude for drop production immediately produces multiple drop ejection or a random spray of drops.
- Tuning, in general, is done by starting at a short pulse width around 1 microsecond, checking to see if there is an amplitude window that can produce monodisperse drops in the desired size range and then increasing the pulse width by about a tenth of a microsecond and checking the drop output as a function of amplitude until stable drops of the desired diameter are produced.
- Drop generators have the size of their drops determined primarily by the size of the ejection aperture nozzle hole. Tuning of the pulse width, however, can produce drops of half the nozzle hole diameter to twice the hole diameter. Most settings produce drops that approximate the aperture hole diameter plus or minus 20%.
- Shifting of the drive amplitude at certain pulse widths can sometimes cause large, abrupt jumps in the diameters of the drops ejected. This is an uncommon phenomenon.
- Changing the pulse width can sometimes effect the directional stability of the ejected drops. Some settings, for difficult-to-understand reasons, eject drops at angles of up to nearly 90 degrees from perpendicular to the ejection aperture. Simply driving the drop ejector at a different pulse width can correct the directional instability and bring the drop ejection direction back to normal.
- The waveform used to excite the piezo element does have an effect on the stability, but the rise time and fall time tolerances for drop ejection to take place are fairly large. Rise times of up to about a couple of microseconds do not seem to affect operation much. Waveforms that have a few microseconds to a few tens of microseconds of fall time seem to decrease the diameter and increase the stability of the drop ejection process when ejecting viscous colloids although very long fall times seem to suppress drop ejection. The ringing of the piezo element makes the waveform, if measured on the element itself, very nonrectangular if excited by a gated voltage source pulse generator. A feedback-controlled linear amplifier allows much better control over the drive process.

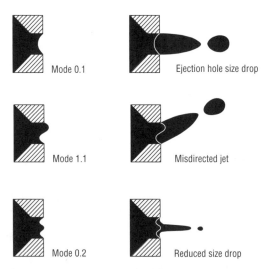

Figure 5.6   The effect of meniscus vibrational modes on drop ejection. The excitation of different vibrational modes in and over ejection nozzle can affect both the size and directional stability of the ejected microdrops.

- One possible mechanism for how altering the pulse width can vary the size of the ejected drop, as well as its directional stability, may lie in the different vibrational modes that ejection pulses with different frequency components set up in the ejection hole fluid meniscus. Examples of how different modes can effect drop production are shown in Figure 5.6. By exciting the drop generator with different pulse widths, we have observed fluid jets that have differences in jet diameters by over a factor of two emerging from the same device using the same fluid. Using more sophisticated manipulation of drive parameters involving sequential excitation of negative- and positive-going pressure pulses, commercial inkjet printers have similarly modulated the diameters of their ejected microdrops.[5,6]

The amount of change in pulse width that droppers will tolerate before they no longer produce stable drops is primarily dependent upon the fluid used. Mixtures of water and propylene glycol had very wide-tuning tolerances. Some of our colloidal mixtures would visibly shift modes of operation with less than a tenth of a microsecond shift in pulse width. In general, what makes fluids ejectable with high stability are:

- viscosity in the 1–30 cS range
- surface tension in the 20–60 dynes/cm range
- low wetting tendency for the aperture plate (can be helped by Teflon coating the aperture)
- low content of particles > 10% of ejection hole diameter

Large-diameter particles that are randomly distributed throughout the working fluid can cause unstable operation by making the break off of the drop from the ejected jet unpredictable since the particle in the fluid can act as an instability node.

5.3.1 Double Pulse Excitation

There is a drop ejector excitation technique in which two closely timed pulses are fed to the drop generator to eject a single drop. The theory is that the first pulse will eject a fluid jet and the second pulse will disrupt the cylindrical fluid column to cause its breakup in a separated drop. The idea behind this is not to rely upon the Rayleigh Taylor instability alone to break up the jet, but to deliberately disrupt the jet at a time that is optimal for the formation of the desired size drop. Double pulse excitation can produce much larger variations in the size of the drops ejected from a given drop generating unit than is possible with single pulse excitation. Also, certain fluid with poor rheological characteristics for monodisperse drop ejection can be ejected stably with double pulse excitation, where single pulse excitation would produce only droplets with accompanying satellites or a random spray.

- Typical values for double pulse excitation would be 1- to 3-microsecond pulse widths and 4- to 30-microsecond pulse separations.
- Changing the separation between pulses has a far stronger effect upon tuning than changing the widths of the pulses.
- The most stable settings for long-term, continuous operation are from pulse separations that are relatively small (i.e., below 20 microseconds). Changes in temperature (which strongly effect fluid viscosity), manometer back pressure, and fluid level, and variations in humidity (which can affect fluid meniscus thickness), can alter the optimal setting for exciting a dropper to produce monodisperse single drops on demand. Different drive settings have different tolerances for these external changes. A way of testing whether one has a fault tolerant drive setting is to see how much one can deviate from the drive parameters and still maintain stable drop ejection. For instance, a setting where one can vary the pulse delays by over 20% of their set values and still maintain stable drop ejection would be a very stable drive setting. Often, one encounters drive settings in which varying the pulse separations by as little as a tenth of a microsecond will stop stable drop ejection.
- Double pulse drive settings are, in general, less stable than single pulse drive settings and in practice much harder with which to find stable operating points.
- Double pulse drive settings usually require about 40% smaller pulse amplitudes to eject comparably sized drops than single pulse drive settings.

5.3.2 Bipolar Pulse Excitation

Bipolar pulse excitation is the use of drive waveforms in which the excursions from the baseline is in both the positive and negative polarities producing positive and negative pressure excursions from the drop generator's quiescent state. This is done in commercial inkjet printers to control the initial firing position of the fluid jet by a controlled withdrawal of the fluid meniscus position. This is also used to control the fluid withdrawal portion of the ejection cycle. In order to obtain higher resolution printing on demand, Epson uses controlled sequential negative- and positive-going pressure impulses to selectively accelerate only the fluid in the central region of the nozzle to alter the diameter of the droplets. The theory behind this method is to initiate a negative-going impulse to draw the fluid in the nozzle inward.

While the fluid is in motion, the fluid is hit with a properly timed positive pressure impulse, causing the central region to be accelerated outward at a higher rate than the fluid near the nozzle walls.[5,6]

5.4 INTERNAL PRESSURE LEVEL

One other aspect important for the operation of a drop-on-demand ejector is the behavior of fluid near the hole in the ejection aperture.

For instance, if the internal pressure is high because of a high hydraulic head or deliberate internal pressurization, the ejection fluid can be forced from the hole in the form of a slow leak. This stops drop-on-demand operation, since jetting through a thick, fluid layer that covers the ejection hole is usually not possible.

Our experience is that negative internal pressurization of the drop ejector reduces the pulse energy needed to eject drops and, in general, increases ejection reliability. We think that this is due to the reduction of fluid meniscus buildup around the outside of the ejection hole. Fluid over the ejection hole acts as an obstacle that the jet must pass through before it can form a free drop. This external fluid also acts as a variable fluidic resistance region that makes the optimal ejection excitation dependent upon the thickness of the fluid layer over the hole. This thickness can be a very strong function of the number of drops previously ejected and the rate at which they were ejected. Fluids that tend to form thick, meniscus films make operation of drop-on-demand devices with random, long, quiescent periods difficult because the optimal pulse pressure amplitudes needed to eject monodisperse drops become a strong function of the prior history of the drop ejector usage. If the fluid meniscus over the hole is asymmetrically located over the hole, the direction that drops are ejected can become unpredictable.

If the fluid's internal pressure is sufficiently high, a continuous stream of fluid can be ejected. This can be used as a basis for fluid ejection in a continuous jet breakup mode in which the piezo element is driven clockwise to produce nodes on the fluid stream. These nodes then form into distinct drops.

Some fluids have a tendency to leak out of the aperture hole. A combination of a low-contact angle with the aperture material, low viscosity and low surface tension can predispose a fluid to have a strong tendency to leak out of the aperture hole and form a thick fluid layer over the outside of the ejection hole. Fluids such as pure water have less of a problem with leaking from ejection apertures. Water has a high surface tension, so it tends to bind to itself rather than wet most surfaces. Water also has a relatively high evaporation rate so that small amounts of leakage in the form of thin films tend to evaporate.

Some fluids when ejected at high drop generation rates leave a residue on the surface of the aperture that builds with time and eventually stops drop ejection. Negative internal pressurization helps this problem by giving the fluid on the outside of the ejection hole a tendency to be pulled back into the dropper. This can be especially important for reliable operation, if the fluid is a carrier for solid particles. If a large build up of this kind of fluid takes place on the aperture and evaporates, it can form a solid deposit over the hole and block ejection.

The level of internal negative pressurization needs to be tuned to the fluid used and the size of the ejection aperture.

High surface tension fluids with high contact angles with the ejection aperture external surface may need no negative pressurization at all to operate. High surface tension fluids, in general, tend not to break their meniscus over the aperture hole and have less of a tendency to start to flow out of the ejector.

Fluids, such as silicone oil that by capillary attraction may form a thin film flowing from the hole over the surface of the aperture, will not pose an operational problem if the fluid evaporates fast enough to prevent macroscopic drop from forming over time, due to fluid accumulation. Observations of low surface tension, low contact angle fluids, such as silicone oils, show they form a micron-thickness film that creeps out of the aperture even with negative pressure applied. Low-vapor pressure fluids that exhibit this behavior require negative pressurization to prevent macroscopic fluid build up on the exterior of the drop generator. It is also important for reliable operation that the fluid leave no solid residue after evaporating.

Drop generators with large-diameter (>50 micron diameters) ejection holes require more precise setting of the negative internal pressurization. Depending upon the fluid equivalent, fluid columns of water of heights as small as 15 cm can cause ingestion of air. This is what limits the amount of negative pressure it is possible to apply. In contrast, we have operated drop ejectors with small-diameter nozzle holes in the 10 micron-diameter range with negative pressures of between 40 to 50 cm of water.

Thin film coating of materials such as Teflon applied to the outside of the ejection aperture can eliminate the need for negative internal pressurization, in cases in which the relative contact angles of the fluids with the thin film coat and materials making up the dropper, eliminate the need for negative internal pressurization.

There are some fluids, such as Ultraol® (Ultra Chemical, Inc., Red Bank, New Jersey) 50 mineral oil (10 cS viscosity), that have very favorable ejection characteristics. However, they also have such a high affinity to glass and silicon surfaces that negative internal reservoir pressure, up to the limit of breaking surface tension, and ingesting bubbles through the ejection aperture is not sufficient to stop a slow leaking out of the fluid. The accumulation of fluid over the ejection aperture hole would ultimately, over several hours, stop drop ejection. One solution adopted in the initial stages of working with this fluid, since the ejection characteristics of this mineral oil was so favorable otherwise, was to orient the drop ejector at a nonvertical orientation so that the fluid that slowly leaked out would not pool over the hole, but would run off to the edge of the drop generator. Application of the thin film Teflon in conjunction with negative internal pressure also reestablished reliable operation. The use of Teflon thin films, however, limited the extent to which the surface could be wiped to remove deposited solids.

5.5 DROP EJECTION RATE

A number of factors limit the maximum and, for some fluids, the minimum pulse rates that are usable for ejecting stable monodisperse drops.

Some fluids deposit a small amount of residue on the outside of the aperture during each ejection cycle. At high rates, this externally deposited fluid can build up faster than it can evaporate away or be drawn by negative internal pressurization back into the dropper. A macroscopic bead of fluid may appear under these circumstances over the hole which in turn stops stable ejection. For some fluids, this limiting drop ejection rate can be in the low hundreds of Hertz.

Even before this macroscopic fluid build up takes place, the fact that a meniscus of varying thickness is being built up over the ejection hole that the fluid must be ejected through, makes the stable generation of microdrops a function of the rate. This produces a situation where the optimal amplitude and pulse width becomes a function of the rate of drop ejection. A significant fluid meniscus build up around the ejection hole makes the process of tuning time dependent. Making a change in the pulse width, for instance, will have a settling time over which the meniscus must adjust to its new equilibrium value. This time period can be on the order of minutes. This greatly complicates the process of tuning to find stable operating points. The key to avoiding this situation is to choose a combination of fluids and surface coatings for the ejection aperture that minimizes this kind of surface wetting.

A meniscus over the ejection aperture hole, which causes a rate-dependent, optimal operating point, can be really damaging to efforts to use a drop-on-demand ejector in which the dropper's duty cycle varies over wide ranges.

The main factor that determines if a dropper has this problem is the fluid used in its relation to its contact angle with the materials making up the ejection aperture. An ideal situation would be for the outside surface of the aperture to have a large contact angle with the working fluid (repelling it), while having a small contact angle with the interior of the dropper (easily wetting it). This can be accomplished by applying Teflon or other similar thin film coatings to the outside surface of the aperture leaving the interior uncoated. Coating of the inside can raise the energy needed to ejection by causing the fluid to fail to wet the aperture so that it must be driven into contact with the surface of the aperture near the ejection hole even before forming an external fluid jet.

One phenomenon that we suspect may produce a rate dependence, is local composition changes of the fluid near the aperture caused by evaporation of volatiles near the aperture and during the fluid jet ejection and withdrawal cycle. It is hypothesized that as the volatile portions of the carrier fluid evaporates, there is a local increase in the viscosity of the fluid near and over the ejection aperture. This local increase in viscosity becomes a function of the ejection rate. A way to minimize this problem, at least for water-based fluids, is to add humectants to the fluid so that the amount of moisture in the fluid gets actively restabilized by absorption from the air and surrounding fluid. Small percentages (5 to 10%) of ethylene glycol, glycerol, and propylene glycol tend to make the ejection of water-based fluid more reliable. This phenomenon is linked with and may be hard to disentangle, in practice, with surface wetting of the ejection aperture over the ejection hole.

At very high rates, another limitation comes into effect, heating of the fluid that changes its viscosity and, if extreme, can depole the piezo drive element and break down or boil the ejection fluid. This comes from both absorption of energy in the fluid and conductive heat transfer from the drive elements.

In addition to having a maximum stable ejection rate, some fluids also have a minimum ejection rate for reliable operation. Fluids, which leave a solid residue when evaporating, harden on contact with air or evaporate off volatiles leaving locally more viscous fluid, have an operating characteristic that, as the interval between droplet ejection increases, the pulse energy required to eject the drops increases. This is known in the inkjet printing literature as "the first drop problem." This is a very important design issue for commercial inkjet printers where the print heads may sit nonoperational for unpredictable periods of time prior to the initiation of a period of droplet ejection. An extreme example of a fluid that would have severe first drop problems would be an air hardening glue. For this type of fluid, any period of time when the fluid is not being actively moved will cause the aperture to be blocked with solidified fluid. There have been many proposed solutions to the problem of rate dependence of pulse ejection amplitude.

Many inkjet printers solve the first drop problem by positioning the print head, prior to initiating a print job, over a special internal cleaning pad and ejecting excess ink at a higher-than-normal pulse amplitude to moisten the surface of the ejection aperture and then to physically wipe the surface of the print head ejection aperture over an internal cleaning pad. Some first generation inkjet, drop-on-demand printers, in fact, had physically removable printheads and an external wiping surface for the user to clean the ejection apertures, prior to each printing job.

A related solution is to make the first drop or series of drops after a period of inactivity be expended into a fluid reservoir prior to depositing drops onto the actual target.

Another solution[7] was to continuously actuate the ejector with pulses that are sufficient only to push the fluid jet out and back, but not sufficient to break off a free drop. When a free drop is desired, the amplitude is increased for that pulse only. This keeps the fluid cycling in constant motion in and out of the ejection aperture. In practice, this may be insufficient for air hardening fluids. We have observed the operation of droplet ejectors that progressively choked off operation requiring higher and higher amplitude actuation pulses over time, even when continuously powered to eject a single drop and finally stopping or throwing out only a random spray of drops. Our experience has been that this approach is best used for fluids that have some marginal dependence on the amount of time between drop ejection pulses, but is not a cure-all for fluids such as air setting adhesives.

Capping the ejector when it is not producing drops is another scheme that has been used by the inkjet industry. Whether this approach is practical depends upon whether one can predict in advance those periods when droplet ejection will not be needed for durations of time larger than that needed to cap and uncap the ejector.

Operating in a solvent vapor saturated environment has been proposed as a way of dealing with the problem of evaporation of solvents from the ejected fluid. The downside is that this can lead to condensation problems unless operated in an environment where the temperature is extremely well controlled.

Various other schemes that implement automated, mechanical, and active cleaning applied intermittently to a stationary drop ejector, or the use of fluid jets to clear debris from printhead surfaces, have been proposed and can be found in a patent database search.

5.6 FLUID FILL LEVEL

If one is using pipette style drop generators with attached piezoelectric drive elements, fluid level can affect the amplitude required to eject drops. In general, if the fluid is above the level of the piezo element, any changes in the fluid level will have a negligible effect of required amplitude. When filled below the level of the piezoelectric drive element, there can be a factor of ten difference in the ejection amplitude needed. This also varies in a nonmonotonic manner. This can be a significant operational complication if one wishes to both automate operation of the microdrop generator and utilize minimum fill volumes. The optimal pulse widths for producing monodisperse drops, however, does not seem to change with changes in fluid level.

Figure 5.7 shows the results of a systematic study on the drive amplitude needed for drop ejection as a function of pulse width and fluid fill level. In addition to the effects of fill level, mechanical resonances in the drop generator produced differences of amplitude required for fluid ejection by factors of two or three common at a given fill level as the drive pulse width was varied. These graphs do not show that the sizes of the drops also varied by a factor of two as pulse width was changed. Different pulse widths also had different levels of satellite drops and amplitude ranges over which monodisperse drop ejection was stable.

The data show that the drive pulse amplitude required for drop ejection is roughly constant with fluid fill level once the fluid level is above that of the piezoelectric element. Increasing the drive pulse width to about 4 microseconds reduces the required pulse amplitude but increasing pulse past this point does not lead to any monotonic decreases in required drive pulse amplitude. If one operates with small fluid volumes that do not fill the drop generator past the piezoelectric drive element, the voltage amplitude required for fluid ejection will increase and decrease as the fluid is used. This is not a monotonic function of fill level. For instance, the drive amplitude required for ejection at a fill level of 5 mm is lower than that required for fill levels of 0.5 mm and 10 mm. Setting the drive amplitude to a level high enough to force ejection at the fluid level requiring the highest amplitude, in practice, usually produces an overdriven spray at the fill levels with a lower ejection threshold.

When we operated drop generators with small quantities of fluid, compensation of the amplitude by hand was possible as the fluid level decreased. In practice, it was not necessary to retune pulse widths or pulse shapes, only the amplitude of the driving pulse. In principle, then, machine vision imaging of the ejected jet can be used to automatically control the appropriate drive amplitude to use as the fluid level drops.

Another method used to stabilize the ejection amplitude of this type of tubular reservoir drop generator when very small quantities of either very rare or very expensive fluids are used, is to prefill the drop generator with an immiscible, non-

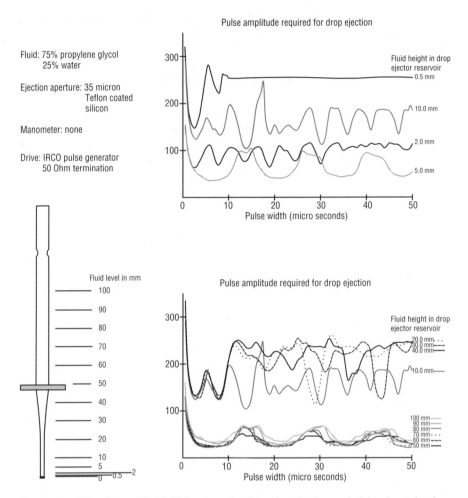

Figure 5.7   The effect of fluid fill level on the drive amplitude needed for drop ejection.
Changes in the electrical drive amplitude need to eject fluid drops as a function
of pulse width and fluid fill level in a Pasteur pipette body microdrop ejector.

reactive, lower-density fluid, such as a light mineral oil, if the rare fluid is aqueous, and then draw in a small volume of the exotic fluid from the tip.

5.7 FLUID TEMPERATURE

We have found that the ejection characteristics of some fluids can be significantly enhanced by operation at elevated temperatures. Changes in temperature of a fluid from room temperature to its boiling point has a minor effect on its surface tension, but a much larger effect on its viscosity. Some inkjet ink components, such as ethylene glycol, can have their viscosity shift by an order of magnitude with a change in temperature of 60°C. Some fluids that will not eject in a stable manner at room temperature will produce reliable monodisperse drops on demand at a slightly elevated temperature. One reason for more reliable operation is the reduction of drive energy required for fluid ejection. High levels of drive can lead to problems associated with cavitating the fluid and ingesting air during the fluid ejection cycle. The reduction in fluid viscosity also enhances operation by suppressing satellite drop formation. This is because high viscosity fluids are prone to form long fluid ligaments connecting the ejection aperture with the ejected fluid drop. Fluid ligaments beyond a certain length tend to fragment at random locations into satellite drops rather than coalesce into the ejected drop or get withdrawn back into the drop ejector. A reduction in viscosity speeds the destabilization of the fluid ligaments so that they break apart at only one point early enough in the ejection cycle so that the two halves of the short fluid jet get drawn into the ejected drop and back into the ejection nozzle opening rather than forming satellite drops. The shift in temperature needed to change an unstable fluid to one that will operate reliably can be small. One fluid consisting of a light mineral oil suspension of powdered meteoric material produced unstable drop ejection at 23°C but stable reliable monodisperse drops at 35°C.

REFERENCES

1. D.B. Bogy and F.E. Talke, Experimental and theoretical study of wave propagation phenomenon in drop-on-demand ink jet devices, *IBM J. of Res. & Dev.*, vol. 28, no. 3, pp. 314–321, 1984.
2. T.P. Theriault, S.C. Winder and R.C. Gamble, Application of ink-jet printing technology to the manufacture of molecular arrays, in *DNA Microarrays — A Practical Approach*, M. Schena, Ed., Oxford University Press, Oxford, 1999, chapter 6.
3. M. Fujino, Progress of color inkjet technology, IS&T/SPIE Conference on Color Imaging: Device-Independent Color, Color Hardcopy, and Graphics Arts V, San Jose, CA, *SPIE,* vol. 3963, pp. 340–348, 2000.
4. J.E. Fromm, Numerical calculation of the fluid dynamics of drop-on-demand jets, *IBM J. Res. & Dev.*, vol. 28, no. 3, pp. 322–333, 1984.
5. S. Sakai, Recording method by ink jet recording apparatus and recording head adapted for said recording method, U.S. Patent 5,933,168, 1999.
6. S. Sakai, Dynamics of piezoelectric inkjet printing systems, IS&Ts NIP16: 2000 International Conference on Digital Printing Technologies, IS&T: The Society for Imaging Science and Technology, Springfield, VA, pp. 15–20, 2000.
7. S. Liker, Methods and apparatus for preventing clogging in ink jet printers, U.S. Patent 5,329,293, 1994.

CHAPTER 6

Troubleshooting Microdrop Ejectors

If one loads up a drop generator with fluid, switches on the drive, and then observes no drops being ejected or that the drop ejection is unstable, there are a number of things to check in order to troubleshoot a malfunctioning drop ejector. This chapter provides a list of aspects of the system to check if drop ejection is either not occurring at all or is happening in an unreliable or suboptimal manner. (See Figure 6.1.)

6.1 CLOGGED NOZZLES

Assuming that the fluid is otherwise well-designed, a dust particle blocking the nozzle ejection hole is the most common cause for a failure to eject fluid. This problem can be minimized by operating with the largest diameter nozzles allowed by the particular application and exercising clean handling procedures for the fluid and the interior of the drop generator.

To test for a blocked nozzle, apply positive pressure and see if a continuous fluid jet, ejected perpendicular to the surface of the ejection aperture, is produced. Applying negative pressure and observing for ingestion of air is less optimal. The first problem is that once air bubbles are introduced, they often are very difficult to eliminate. The other problem is that clogged apertures can be capable of ingesting air under negative pressure but unable to eject drops or push out bulk fluid under pressure. On many occasions we have observed a large particle resting over the nozzle hole on the inside of the ejector acting as a one-way valve.

Drop ejectors using glass reservoir tubes can be visually inspected for particles prior to filling by backlighting them and observing them at moderate magnification with a binocular inspection microscope. Observing down the bore of the drop ejector with a microscope and checking to see if the hole is clear is sometimes possible but, in general, a less reliable method of checking for a blocked nozzle. The technique

71

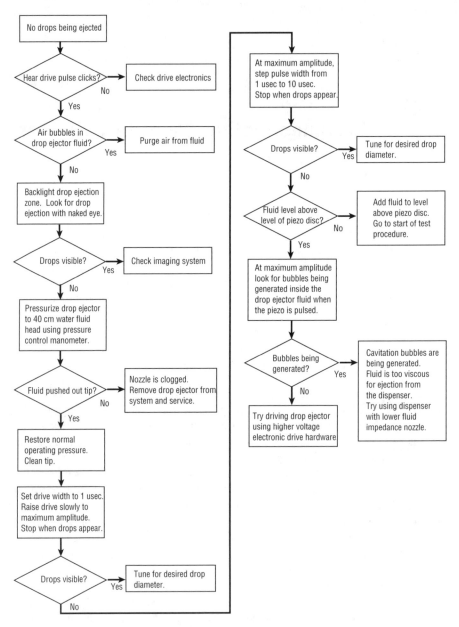

Figure 6.1   Troubleshooting flow chart for a drop dispenser that is failing to eject microdrops.

is useless when using opaque fluids. In addition, we have used organic fluids that formed transparent hard thin films that blocked off fluid flow on contact with air.

One cannot overemphasize the importance of cleanliness and particle control in maintaining reliable drop ejector operation. In addition to blocking off the nozzle totally, particles lodging in or over the ejection nozzle hole can make operation very erratic. A drop ejector that abruptly changes its drop ejection characteristics may have a partially blocked ejection hole. Fibrous dust particles lodged in the ejection nozzle can still allow drop ejection but seriously interfere with drop ejection stability. Cleaning a drop dispenser prior to use is a very important part of normal operating procedure.

The preferred procedure for cleaning a newly assembled drop generator depends upon the diameter of the ejection aperture hole. For ejection holes larger than 20 microns we attach a vacuum line to the fluid reservoir fill port and immerse the tip in distilled particle filter water. The cleaning fluid is drawn by vacuum in through the nozzle. This acts as a further guard against introducing particles that are larger in diameter than the ejection hole into the interior of the dispenser. After the drop ejector is filled, the water is removed by pressurizing the drop ejector by blowing into the tip with the spray from a can of dust removal compressed gas and jetting the water out the fluid fill side of the reservoir. We repeat this cleaning cycle until examination of the fluid reservoir by an inspection microscope shows no visible particles on the interior of the ejector. The fill port is then plugged with a rolled up piece of lint-free optical cleaning paper. Freedom from particles that are capable of clogging the tip is verified by drawing in a quantity of particle filtered distilled water by dipping the tip into the fluid with the interior of the drop ejector under vacuum and then jetting the water out the tip by applying positive pressure.

Drop ejectors with ejection aperture holes smaller than 20 microns in diameter and with large fluid reservoir volumes are impractical to flush by drawing in fluid from the tip under a vacuum. It simply takes too much time to fill. For these small diameter nozzle drop dispensers, a syringe injects the particle filtered flush water via a length of small gauge Teflon tubing into the drop dispenser fluid reservoir. The tip of the Teflon tube is placed all the way down to the tip of the drop ejector tube and pressurized particle filtered distilled water is jetted into the drop dispenser and allowed to flow out the back of the dropper assembly. To insure that the tip is flushed, pressurized filtered air is directed from the outside of the dropper into the ejection hole. This will push out any fluid plug that may have formed by surface tension in the tip. A vacuum pump can be applied to the drop ejector to assist in evaporating out any residual cleaning water. After the cleaning operation is completed, a plug composed of rolled up, lint-free optical paper is used to seal the open end of the drop ejector. This allows air flow to facilitate filling and establish the necessary operational internal pressure levels while blocking the introduction of particles.

Solvents such as acetone and methanol can also be drawn in and expelled from the tip to clean the interior of the drop generator. It may be necessary to use such solvents in cleaning out a drop ejector that previously was used with nonaqueous fluids. Using vacuum and pressure to fill and empty from the tip and keeping the back of the drop generator sealed with a gas permeable plug helps keep dust particles out of the interior of the drop generator.

6.2 DAMAGED NOZZLES

We have also observed drop generator malfunctions that have been caused by physically damaged nozzles. Both ground glass and micromachined nozzles can crack out sections adjacent to the ejection hole producing an enlarged, irregular-shaped ejection aperture hole. The main operational symptoms of this are the ejection of abnormally large drops regardless of drive settings and ingestion of air at very low negative pressures. This kind of damage is readily confirmed by direct observation of the exterior of the ejection aperture with a microscope. Small, cracked-out regions can lead to misaligned or secondary small jets and directionally unstable ejection. If the drop generator seems to throw misdirected jets, regardless of drive settings or attempts to clean the aperture surface, one should check for this type of damage.

One way the ejection apertures can be damaged is during ultrasonic cleaning. The use of ultrasonic cleaners is a very effective way of unclogging jammed drop ejectors, but when done improperly, can seriously damage the ejection aperture. If the surface of the ejection aperture comes into direct contact with the walls of the ultrasonic cleaning vessels, the chances of damage occurring is very high. We hypothesize that this is because the two solid surfaces of the wall of the cleaning bath and the ejector aperture coming in direct contact mechanically hammer each other. Immersing the drop generator in the fluid of the ultrasonic bath, in our experience, has not caused damage to an aperture plate able to withstand the stress of drop ejection.

6.3 INTERNAL AIR BUBBLES

Air bubbles in the fluid near the ejection aperture can adversely affect fluid ejection by acting as hydraulic shock absorbers and increasing the drive amplitudes needed to eject fluid drops. One of the advantages of glass body microdrop ejectors is that these ejection destabilizing air bubbles can be visually detected readily.

These bubbles can be introduced into the fluid in a number of different ways:

- Air bubbles can be introduced into the fluid during the filling process.
- Air bubbles can be ingested into the drop generator by an excessively high negative pressure. Some negative pressure is needed for most fluids to prevent the fluid from slowly leaking out the ejection aperture hole. Excessive negative pressure can cause air ingestion. The amount of negative pressure it is possible to apply without compromising operation varies strongly with the diameter of the ejection hole. Large holes, 75-micron diameter and up, can break meniscus and ingest air at fluid heads of as low as a –10 cm column of water, depending upon the operating fluid.
- Air bubbles can be generated by internal cavitation of the fluid from high-drive pulse amplitudes. When attempting to clear out clogged ejection apertures by using high amplitude drive pulses, bubble generation in the fluid in the ejector reservoir have been observed. Once generated, these bubbles do not redissolve in time frames short enough to permit continued reliable operation. This fact implies that one cannot arbitrarily increase the ejection pulse pressure amplitude to eject highly viscous fluids. Ultimately, the pressure impulse reaches an amplitude to

where the fluid cavitates and starts to build up internal air bubbles. For some fluids, the threshold of cavitation can be low enough to effect near normal operating conditions. We have observed erratic operation of some drop generators where one of the principle causes was an operating fluid that had a threshold of cavitation that was just slightly above that of the threshold of ejection. Slight missteps in tuning up the drop generator would result in bubble generation and internal gas bubble build up near the ejection nozzle.

- Air bubbles can be ingested as part of the fluid ejection cycle. We have observed this effect when using metal foil pinhole apertures as fluid ejection nozzles. The holes were large in diameter in comparison to the foil thickness. This allowed air drawn in to collect on the interior of the drop ejector rather than to be pushed out by capillary action refill, as in the case for conical profile holes or longer cylindrical cross section holes.

6.4 MISDIRECTED JETS

One undesirable mode of operation is where the drops are not ejected perpendicularly to the ejection aperture face but at an angle. This can cause problems in the predictability of the trajectory of the drops, since the angle of misdirected jets usually varies with the ejection speed of the drop. There are a number of causes of misdirected drop ejection:

- Nozzle damage
- Partially clogged ejection holes
- Asymmetric wetting over the ejection hole
- Debris build up over the ejection hole
- Mistuned drive

The first four causes of misdirected jets basically have the same physical mechanism. A fluid jet that contacts any surface, either fluid or solid, gets pulled in that direction by an amount proportional to the time in contact to the defect and will suffer a proportional amount of deflection. The effects of an asymmetric meniscus can be dramatic. Ejected drops that traveled nearly parallel to the flat surface of the aperture plate have been seen. The effects of small, micron-dimension nozzle defects on the directionality of ejected drops was quantified by studies done by R.E. Drews at Xerox in the 1991 paper "The effects of translationally symmetric nozzle face defects on the directional accuracy of thermal inkjet arrays."[1] The research at Xerox found that step heights asymmetries around the ejection hole on the order of microns could cause deflections on the order of 30 degrees. For a given surface defect, the deflection decreased as the ejection speed increased.

We have found that the design of the ejection aperture affects the tendency for drop ejectors to have misdirected jets. Micromachined structures that have acute angle, sharp-edged aperture holes have had a greater tendency to throw misdirected jets than those that terminate in a straight cylindrical section. We speculate that this is because a sharp-edged, acute-angle hole does not easily define a consistent meniscus position or a preferential direction of fluid flow, as well as a short, right-angle cylinder. Sharp edge holes are also easily damaged by cleaning and physical contact leading to asymmetric wetting around the edges of the hole. Ejection apertures that have a planar outer surface terminating in a short, right-angle cylindrical hole have the practical advantage that they can be physically wiped to remove debris. This often restores proper ejection. Ground glass tip ejection apertures are much more likely to suffer damage from this type of cleaning process.

Sometimes one may encounter behavior where certain drive settings produce misdirected jets while others produce proper perpendicular ejections. One explanation is that the drive impulses have frequency components that are exciting asymmetric vibrational modes in the fluid meniscus. Figure 6.2 shows how this might cause a drive impulse dependent misdirection of the ejected drops.

In order to avoid deflected jets, one needs to design the ejection aperture structure to be as intrinsically axially symmetric as possible. The interface between the interior of the ejection channel and the outside surface should allow a well-defined location for the fluid meniscus surface when the fluid is under negative pressure. Here, a right-angled cylinder or an aperture with an inner step would be better than a tapered cone or pyramid terminating in a sharp-edged hole. Ideally the surface of the ejection aperture should be kept dry by appropriate engineering of the fluid and the contact angle of the exterior surface of the nozzle. If this is not practical, the surface should be physically uniform enough that the fluid build up is symmetric around the ejection hole. Some provision for periodically wiping the surface should be made so that any asymmetric buildups caused by dust or the irregular deposition and evaporation of the fluid can be remedied. Ground glass tip ejectors that cannot be wiped without significant risk of damaging the surface can sometimes be cleaned by immersion into a solvent under an ultrasonic agitation. There are some fluids, though, where this cleaning process cannot be used without damaging the fluid payload, such as large biological molecules or living organisms. In cases where there may be irregular build up of particulates that may cause misdirected jets, the ejection speed of the drops should be as fast as possible without causing unacceptable deterioration of the integrity of the ejected drops. This can bring an otherwise unacceptable level of deflection down to a level that is within operating specifications. One should also experiment with different tuning settings for the drive impulses to insure that one is not operating at a pathological operating point that is exciting asymmetric standing waves on the meniscus surface.

6.5 OVERLY WETTED MENISCUS

A slight asymmetric wetting of the surface over the ejection hole can cause significant deflections of the ejected fluid jets. A thicker fluid layer, even if symmetric, can raise the pulse energy required to eject drops or suppress drop ejection entirely. The slow leaking out of fluid from the ejection aperture hole is normally suppressed by a combination of negative internal pressure and the application of a high contact angle surface to the exterior surface of the ejection aperture. In some cases, both are needed. We have used certain light mineral oils in which negative

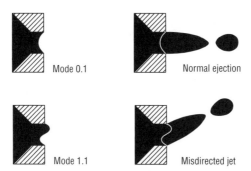

Figure 6.2 Possible mechanism for why variations in the drop ejector drive can affect the directional stability of ejected microdrops.

internal pressure up to the point of drawing in air was insufficient to stop the slow leaking out, due to surface wetting of the oil. The application of both a Teflon-like fluorocarbon thin film and negative internal pressure were needed to prevent a cumulative fluid build up that stopped fluid ejection in a matter of hours. This was a rather pathologically bad fluid, however. Another mechanism for building up unwanted fluid is electrostatic attraction of ejected drops back to the aperture plate. This can be a particularly serious problem when ejecting small drops that come to terminal velocity less than a mm away from the ejection aperture plate. Drops can come out of ejection apertures charged to tens of thousands of electron charges or more. This is sufficient to have the drops drawn back towards the aperture plate rather than falling free. The same process can occur with small satellite drops. If electrostatic-induced wetting is a serious problem for the fluid used, the charges that drops assume when they are ejected can be controlled by grounding the fluid and using an electric field generating loop through which the drops are ejected.

6.6 INTERNAL PRESSURE

Misset internal pressure can disable drop production but is easy to detect. Too negative an internal pressure will draw in air bubbles. Insufficient negative pressure can cause fluid buildup over the meniscus hole.

Large-diameter holes are more difficult to set operating pressures for than small-diameter holes. The effect of surface tension in keeping a fluid meniscus in place against gravity and internal-external pressure differentials makes apertures with ejection holes in the 100-micron diameter range subject to ingesting air with as little as a 10-cm column of water pressure differential. For some of these fluids, contact angle modification of the exterior of the drop generator is an essential part of keeping the fluid contained due to the limitation of the amount of negative pressure that it was possible to utilize without drawing in air. Over the range of internal pressures that allows ejection of drops, varying the internal pressure will change the drive pulse energy that is optimal for ejection of drops. The relationship that we found was nonintuitive, though. With increasing negative pressure, the drive amplitude needed to eject drops decreases. As a consequence, for fluids that have a narrow range of drive amplitudes where monodisperse drops are ejected, stabilizing the internal pressure against drifting can be essential to insure reliable operation.

6.7 INSUFFICIENT DRIVE AMPLITUDE

As the viscosity of the fluid increases, so does the amplitude of the drive needed to eject that fluid. Past 100 cS, the drive amplitude needed to eject microdrops may be beyond what can be used without damaging the drop ejector or causing cavitation in the fluid. If driven hard enough, the piezoelectric drive elements are capable of breaking the glass reservoir tubes to which they are attached. However, this requires a rather extreme overdrive factor. As an example, one glass reservoir tubular drop generator that would eject water drops at a 20-volt drive level was cracked in half during testing by a drive pulse amplitude of 2000 volts.

One can verify that drive impulses are reaching piezoelectrically driven micro-drop ejector by listening for the clicks that are produced when the drive electronics power the piezodrive element. Stroboscopic imaging can be used to see if a jet is being formed with insufficient energy to break off and release a free microdrop. Ultimately, the internal pressure levels are hard limited by cavitating the fluid, breaking the drop ejector, or depoling the piezoelectric drive element.

If these limits are not being pushed, and one is limited by insufficient drive from the pulse source, there are a number of things to try. The first is to systematically try varying the pulse widths going from 1 to 100 microseconds to see if there is a setting that will produce drop ejection. We have seen the effects of varying the pulse width on the threshold of ejection for a given drop generator fluid combination to be on the order of two to one. If the drive electronics will allow this mode of operation, one can try to use pulse pairs to drive the piezoelectric element. In general, the use of pulse pairs will eject drops with a reduction by about 50% from single pulse settings. Typical pulse pair settings would be pulse widths of between 1 to 5 microseconds and pulse separation of from 5 to 20 microseconds. If the output stage of the pulse source can drive an inductive load, a pulse transformer can be used to increase the amplitude of the output drive. Not all pulse sources have a low enough output impedance and enough stability to operate in this mode, however. If varying the drive within the limits of fluid cavitation and drop ejector hardware limits cannot produce controllable fluid jets, in some cases heating the tip of the drop ejector can lower the viscosity sufficiently to eject microdrops with an otherwise unejectable fluid (Section 5.7).

6.8 NONOPTIMAL PULSE WAVEFORM

A nonoptimal drive waveform can be the cause of nonejection, ejection of misdirected drops, and ejection of drops with unwanted accompanying satellite drops. While some fluids are more tolerant of nonoptimal drive pulse waveforms than others, it is always good practice to systematically vary the drive waveform to see if poor operating performance is due to a pathologically bad fluid or a poor drive setting. We usually start by using single-pulse drive settings of 1- to 10-microseconds width to find the operating point that yields drops in the size range desired, which are free from satellites with the widest operating voltage window between the threshold of ejection and the destabilization amplitude. If no settings are found using single-pulse settings, then double-pulse settings and modification of the rise and fall time of the drive waveforms are attempted. (See also Section 6.4, Misdirected Jets and Section 6.7, Insufficient Drive Amplitude.)

6.9 PULSE RATE

Some fluids drop generator combinations have stable ejection only within a narrow range of ejection rates. There are multiple causes for the rate at which drops are ejected to affect the reliability of drop ejection.

A very viscous fluid limits the rate of capillary action refill of the ejection aperture.

A fluid that leaves residues on the exterior of the ejection aperture as part of the ejection cycle requires time for the negative internal pressure to draw the external fluid back into the drop ejector. For this type of fluid, too high a rate can result in a build up of fluid over the ejection hole, which can stop drop ejection.

Too low a rate, and one can run into the "first drop problem" in which the evaporation of the fluid in or over the drop ejection hole changes its rheologic properties such that a different, usually high, ejection pulse amplitude is needed to eject a drop as compared to a higher repetition rate where the composition of the liquid over the ejection hole is that of the bulk fluid.

Mechanical resonances in the drop ejector can be excited leading to specific operating frequencies where the size of the drop or its directional stability abruptly changes.

6.10 PROBLEMATIC FLUIDS

Not all fluids can be jetted from drop-on-demand microdrop generators to form monodisperse drops. There are some common properties that can stop fluids from ejecting. Problematic fluids are fluids that:

- Have excessively high viscosity (> 100 cS).
- Contain particles large enough in diameter to block the ejection aperture hole.
- Contain particles that settle onto the bottom of the drop ejector, covering over the hole with "mud."
- Contain particles that agglomerate together over time into clog-forming, large masses.
- Can evaporate leaving a solid skin over the hole or a solid or high-viscosity fluid plug in the hole.
- Contain live organisms that form clumped colonies over or in the ejection hole.

- Wet surfaces so strongly that they rapidly leak out the ejection hole blocking the outside of the ejection hole regardless of applied negative pressure.
- Have such a high-contact angle with respect to the inside of the drop ejector that they internally bead up and fail to wet the inside surfaces of the ejection aperture.

Some fluids will jet but will not form monodisperse drops. What is observed can be drops always accompanied by satellites or a chaotic spray of drops. Some fluid properties that can cause this type of behavior are:

- Excessively low surface tension.
- Contains particles or particle agglomerates with diameters approaching that of the fluid jet.
- Inappropriate surface tension relative to fluid viscosity.

The easiest way to check if the fluid and not the drop generating hardware is the problem is to replace the fluid with one that is known to easily form stable, monodisperse drops. The best fluids we have found for this purpose are particle filtered mixtures of propylene glycol and distilled water. A 75% propylene glycol, 25% distilled water mix was able to form monodisperse drops in all drop generators we tested.

6.11 MECHANICAL MOUNTING

Erratic operation of a microdrop generator can be due to the mechanical mounting not being consistent. This is particularly a problem with long, tubular fluid reservoir drop ejectors. Long glass structures excited at a single point can have mechanical resonances excited that can couple back into the fluid. Since mechanical resonances excited in the drop generator can affect the drop ejection process, a drop ejector that is loosely mounted and shifting position will require retuning based on the contact it is making with its holder at any given point in time. This effect is easy to observe if one is holding a drop generator in one's hand and is watching the drops being ejected with a strong backlight. As one's fingers touch the drop ejector in different locations, different drive voltages will be needed to keep the drop ejection stable.

6.12 IMAGING SYSTEM

If the ejection of drops cannot be observed, one possible reason is that the imaging system is not set up properly. It is very possible to eject drops and, because of an improperly setup imaging system, not to see them.

When using a strobe illumination imaging system and a synchronized camera, the drop generator can either be operating synchronously or asynchronously with the camera. Operating synchronously with the camera is often not possible due to the frame rate limitations of most digital cameras and image capture cards. The microdrops can be traveling as fast as tens of meters per second near the ejection aperture. At high magnifications, the time fraction during which a drop will be in the field of view of the camera can be small. For instance, a typical real space field of view might be 1 mm square.

A typical imaging rate might be 10 Hz. If one is ejecting one drop per second at 10 meters per second asynchronously with the imaging system frame rate, any given drop after ejection will be in the camera field of view for only 1/10,000 of a second.

At a strobed illumination imaging rate of 10 frames per second, it is very easy to see how the drops can pass through the camera's field of view and not be imaged. Operating asynchronously, one increases the chances of imaging the microdrops if one images them at a point in space where they have decelerated to their terminal velocity, which, depending upon size, is on the order of a few mm to a few cm per second. If the drop's dwell time in the field of view is longer than the reciprocal of the frame rate, then the drop will be imaged in each frame even if the camera and the drop ejector is operating asynchronously. Operating at a high drop rate as is practical will also reduce the chance of not producing drops and having the imaging system not pick them up.

Operating synchronously, in principle, one can image a high-speed drop every frame if the timing is set properly between the drop ejector and the imaging system. To avoid missing drops due to a timing error in setting relative delays, one can synchronize the drop ejection and the imaging in a synchronously operating drop imaging system by setting the drop ejector to overdrive. This will produce satellites and spray that will have widely varying speeds and trajectories and will then appear in many more frames than a monodisperse drop stream. These satellite and spray drops will furnish image references that can be used to alter the relative phasing between the drop ejection and the imaging.

Backlighting the drops with a bright collimated light source and viewing the drops against a dark background with the naked eye can be used as an auxiliary check on electronic imaging systems. Under the right lighting conditions, sub-10-micron diameter drops are easily visible. The key to seeing small drops with this method is to view the forward scattered light by using a highly collimated beam and observing the drops with one's eyes facing towards the beam as much as possible and blocking off the light from directly entering the eyes using a light shield. The drops will be invisible, if one tries to view them at any angle but that facing the light source.

6.13 RECOMMENDED STARTUP PROCEDURE TO TEST A NEW DROP GENERATION AND IMAGING SYSTEM

- Use a fluid known for good ejection characteristics such as a particle filtered mixtures of 75% propylene glycol and 25% distilled water.
- Use an aperture size of 20 to 50 microns.
- Briefly apply positive pressure to force out a small amount of fluid to make sure that the ejection aperture hole is clear.
- Reset to an internal pressure of -10 cm of water for the actual test.

- Physically wipe the surface of the aperture to insure that there is no excess fluid on the surface.
- Set the ejection rate to 100 Hz for ease of drop tracking and listening for piezo-actuation.
- Set the drive electronics for a 10 microsecond pulse width and listen for actuation of the piezodrive element in order to verify proper operation of the electronics.
- Use backlighting by a bright beamed illumination source to initially view the drops by eye in addition to using the electronic machine vision imaging system.
- Use a single pulse actuation of the piezoelectric drive element of between 1.0 to 10.0 microseconds using trial and error to find the optimal pulse width.

REFERENCE

1. R.E. Drews, The effects of translationally symmetric nozzle face defects on the directional accuracy of thermal inkjet arrays, IS&T's Seventh International Congress on Advances in Non-Impact Printing Technologies, pp. 107–116, IS&T: The Society for Imaging Science and Technology, 1991.

CHAPTER **7**

Imaging Microdrops

Microdrops are commonly imaged by three basic methods:

- Directly illuminating the drops using a constant, bright-beamed light source and observing the drops as points of light against a dark background with the naked eye or with optical instruments such as remote magnifiers.
- Creating a bright, uniform background with an illuminated diffuse ground glass screen and viewing the drops as dark shadows against a bright background using an electronic stroboscopic imaging system.
- Direct illumination of the drop using a strobed light source and observing drops as bright objects with a machine vision system.

7.1 DIRECT VIEWING OF ILLUMINATED MICRODROPS

A strong collimated backlight is the simplest and least costly of drop imaging systems.

Backlit against a dark background, the naked eye can be used to observe drops as small as an estimated 10 microns in diameter. Optimally, the light source should minimize the heating of the microdrops and their environment. Examples of light sources that provide high brightness with minimal heating are lasers, high brightness light emitting diodes (LED), and fiber-bundle-coupled halogen lamps. Hot light sources, such as collimated incandescent sources, are often not usable because they heat the fluid in the drop ejector, changing its rheological properties, and generate strong, local convection currents in the air that will deflect the drop's trajectories. The drops are most easily viewed from a position where the illumination beam is shining towards the observer as much as is possible without being in the observer's eyes. The forward, scattered light from the small drops is significantly brighter than the reflections from the rear or the side of the drops.

Figure 7.1   35-micron diameter fluid drops ejected into free room air, back-illuminated by a fiber-optic-coupled halogen light source.

Backlighting the drop ejector to view the ejected fluid has the advantages of simplicity, low cost and ease of interpretation. Electronic imaging systems utilizing strobed sources and frame grabbers can sometimes fail to register that drops are being ejected. This can occur if the ejection and the strobing are misphased such that the drops have fallen out of the field of view before they can be imaged. Another way that electronic imaging systems can fail to pick up ejected drops is if imaging is being done near the ejection aperture where the velocity of the drops is very high. This high-velocity region may extend out past 1 or 2 cm for large-diameter drops. Unless the ejection rate is very high or the strobe is properly phased, it is very easy to be ejecting drops and not see them. During the initial troubleshooting phase of characterizing a fluid mixture for a particular drop ejector where there is a question about whether the fluid recipe will eject at all, a backlight setup is a very quick and easy way to evaluate fluid ejectability (Figure 7.1).

7.1.1   Backlight Illumination Sources

The key is to be able to be able to illuminate the drops with a bright beam that is directed towards the viewer while simultaneously having a dark background against which to view the drops. This requires an intense source of light capable of being directed into a beam with a sharp edge cutoff. It is also highly desirable that the illumination source not be a source of excessive heat that can significantly alter the temperature of the fluid in the drop ejector and set up convection currents.

Ultra-bright LEDs need light blocking tubes to avoid dazzling the eyes, rendering the drops invisible. Red and orange LEDs function marginally for visual use but are adequate for machine vision systems that have better relative sensitivity in the red than human vision. The drop ejector had to be within 1 or 2 cm of the LED for the drops to be visible to the naked eye. The recent new generation of gallium nitride high brightness short wavelength LEDs function far better. The latest generation of shorter wavelength LEDs presently being used in LED flashlights has very practical droplet illumination elements for naked eye microdrop viewing.

A 5 milliwatt (legal safe limit) laser with a controllable defocusing lens to vary the size of the spot works quite well if the drops can be kept within the laser beam. This performs far better than the LED because the contrast ratio between the background and the illuminated drop is very high. SLAC researchers have used this method of drop viewing to center streams of drops into submillimeter mechanical apertures. The local heating from a low-power laser did not disturb the fall path of the drop in stagnant air (the experiment was in a convection-controlled chamber), as would a more powerful conventional light source such as a halogen lamp.

The exact setup for this alignment routine used a small, close-focusing telescope to view the illuminated drops, which were within a double-walled, convection-shielded chamber. The laser was aimed towards the telescope with a 10-degree offset. A piece of tape on the window facing the telescope was used as a beam block.

The halogen lamp works because it offers a high-surface brightness, concentrated illumination source that, unlike conventional incandescent sources, can be shielded in a way to provide a collimated high brightness beam with which to illuminate the drops against a dark background. The big advantage of halogen lamps is the large area of illumination over that provided by eye-safe lasers. The major disadvantage is the high heat produced by most halogen sources. The best type of halogen illumination source we found was a fiber optic coupled halogen lamp. The fiber bundle transmitting the light eliminated a large fraction of the nonvisible wavelengths reducing the heat problems. The fiber optic light transmitting cable also allowed the illuminator to be introduced to more optimal points for drop viewing. A convenient field troubleshooting illumination source for drop viewing is an ultra bright lithium battery, halogen lamp pocket flashlight such as the SureFire® (18300 Mount Baldy Circle, Fountain Valley, CA) line of ultra bright flashlights.

7.2 BRIGHT BACKGROUND IMAGING OF MICRODROPS

Background illumination using a stroboscopic light source against a ground glass screen combined with electronic imaging with a synchronized camera and frame capture device is the most common method used in the inkjet literature for imaging microdrops. The advantage is that heating of the drop by the illumination source is minimized as opposed to those schemes that require making the drops optically brighter than the surrounding background. An additional reason for uniform bright background imaging is that when using computer imaging hardware it is much easier to make a uniformly illuminated background and image the drop as a shadow than to use direct illumination of the drops to make the drops bright sources of light while retaining sufficient, nonsaturated dynamic range to see large primary drop along side, small satellite drops.

Short duration, high intensity, pulsed illumination is used so that the drops can be discretely imaged as rounded objects rather than as streaks, and so that the complex structure of the high-speed fluid jet can be seen. This is important in detecting satellite drops and in studying the behavior of the ejection jet as the fluid,

Figure 7.2 Bright background shadow image of a 35-micron diameter microdrop forming from a drop-on-demand device.

which is ejected from the ejection aperture hole, is formed into free microdrops. These small scale fluid jet structures are far better imaged in practice by viewing them as dark objects against a bright background rather than as illuminated bright structures. Viewing ejected microdrops as dark objects shadowing a bright background is the method used in nearly all published studies of microdrop generation.

Uniform illumination is important for many scientific applications where the drop must be detected and centroided using machine vision algorithms. Uneven illumination would throw off the apparent center of mass. The uniform background is typically produced using a ground glass screen illuminated either by a gas discharge strobe or an array of pulsed-light-emitting diodes (Figure 7.2).

For studying drops at terminal velocity, a pulse duration of a few tens of microseconds is adequate, but for studying the ejection process near the jet, the pulse width must be of durations on the order of 10 microseconds or less. As an example, the initial ejection speed of a 15-micron diameter drop is about a meter/sec, while its terminal velocity in air is about 7 mm/sec. At terminal velocity, the 15-micron diameter drop would move 0.7 microns allowing acceptable imaging. However, a 100-microsecond pulse width would allow for about 100 microns of movement at its initial ejection speed rendering phenomenon occurring at the initial ejection jet unimagable. A 1 microsecond flash from a Xenon strobe, on the other hand, would only have a drop movement of 1 micron in the high-speed regime just after ejection. The effective pulse width of gas discharge tube strobes is on the order of a microsecond.

Until recently, LED arrays had to be pulsed on for durations of tens of microseconds to produce a usable image due a combination of the LED's relative inefficiency and the lack of sensitivity of available cameras. For this reason gas discharge strobes were the only practical means of providing stroboscopic illumination of high speed fluid jets. The practical disadvantages of gas strobe illumination are high cost and poor flash-to-flash reproducibility.

Starting at $2000 with replacement strobe bulbs costing in the hundreds of dollars and lasting for only a few months with continuous use, commercial strobes suitable for use in machine vision imaging are also expensive sources. In contrast an LED based illuminator can be constructed from common parts for less that $100 and lasts indefinitely. Since the strobe flash is an electrical discharge between electrodes in a closed bulb, every flash is slightly different in intensity and light distribution from every other flash. This is due to erosion of the electrodes and deposition of vaporized material on the inside of the bulb. The LED array however will have a consistently reproducible flash-to-flash illumination profile. LED arrays since 2001 have become competitive with gas strobe sources for microsecond interval stroboscopic illuminators due to the steady increase in LED efficiency and the increasing sensitivity of solid state cameras. It is likely that for microsecond class stroboscopic imaging gas strobe sources will shortly be made obsolete.

7.3 BRIGHT SOURCE STROBOSCOPICALLY ILLUMINATED DROPLETS

Viewing drops as dark shadows against a uniform bright background, while the best way to study drop trajectories and the process of drop ejection quantitatively, may not be a practical way to perform machine vision imaging of microdrops when attempting to direct a stream of microdroplets into a targeted structure. This is because the geometry of drop ejection assemblies intended for manufacturing often does not permit one to place a bright, uniformly strobe illuminated background behind the drops at all important points in its trajectory from the point of ejection to the impact of the drop at its target.

There is an alternate way to view the trajectories of ejected microdrops using a machine vision system. By taking a stroboscopic illumination source and illuminating the ejection path of the droplets with the illuminating source shining toward the camera, as much as is practical, the ejected droplets will appear as bright spots superimposed over the image of the drop ejector and the ejection target. As much as a 45-degree angle between the direction of illumination and the viewing axis is tolerable, though the closer the illumination source is to shining straight towards the camera, the brighter the drops will appear. The instantaneous brightness level of the source required is on the same order as required to back illuminate the drops when viewing them as dark shadows against a backlit screen. A stroboscopic source is used so that the drops will appear as points of light rather than streaks, as would be the case for continuous beam illumination (Figure 7.3).

One additional advantage of using bright droplet viewing techniques is that at a given magnification, by using high-brightness, narrow-beam illumination sources, the smaller droplets can be made more visible than when using bright background viewing. There is a lower limit to the size of an object that can be resolved in bright background techniques, which makes the droplets visible by having the droplets physically block existing background light. The fundamental advantage of bright source viewing in making very small droplets visible is that there is no corresponding limit to how small an object, such as a droplet, can be viewed if the object is self luminous.

The disadvantage of this technique over viewing the drops as dark shadows against a uniform bright background is that the details of the ejection jet is very poorly resolved compared with bright background shadow imaging. Compared with bright background viewing, precise microdrop positional centroiding is far more difficult since it is required that the illuminating beam be made extremely spatially uniform.

Figure 7.3   Droplet stream made visible by being illuminated with a synchronized strobed LED array. The microdrop ejector in this figure is being imaged ejecting 35-micron diameter drops next to the edge of a dime.

All published photographs in the literature detailing the droplet jetting and drop formation process, view the jet and drops as dark objects against a bright background.

7.4 LIGHT EMITTING DIODE ILLUMINATION ARRAYS

Obtaining maximum pulsed optical output from LEDs is a bit of an art since operation of LEDs in very low-duty cycle, high-current pulses on the order of tens of microseconds is something not typically specified on manufacturer's data sheets. LEDs operated at 100 microsecond pulse widths or less can have peak currents with corresponding high optical outputs a factor of 20- to 50-times higher than the maximum continuous operation limit. Ultimately, there is an upper-current limit based not on integrated average dissipated power but on internal junction current density limits. Past this limit, the LED may fail in one pulse due to a debonded contact wire, or slowly degrade over a period of hours to days.

One sign of this limit is where increases in the pulse current do not yield proportionate increases in the optical output. Researchers at SLAC tested LEDs for maximum output by having them illuminate a high-speed optical receiver and raising the pulse current while at the operating conditions anticipated (100 microsecond pulse width, 10 Hz pulse repetition rate). They then raised the pulse current in steps graphing the relationship between the measured optical pulse amplitude and the drive current. One caution is not to assume that the voltage drop across the LED is constant when operated in this mode. The nominal 1.4 volts across a gallium arsenide LED, when operating at a few tens of milliamps, is not what will be measured when driving it at between one and two amps. Voltages as high as 3.5 volts are measured across a pulsed LED. Once a break point is seen where the increase in current does not produce a proportionate increase in optical output, the operating current is cut back to about 20% and an extended duration test lasting several days is conducted to ensure that this current level does not cause long-term degradation of the LED. This kind of testing is time consuming but pays off in significantly decreasing the number of LEDs required in an illumination array for a given desired pulse width.

At SLAC, researchers have found that manufacturers' test conditions for pulse operation do not cover the low-frequency short pulse widths that are useful for observing microdrops and significantly underestimates the peak power that an LED can produce. For instance, the Agilent Technologies HLMP-EG08 SunPower® (Agilent Technologies, 395 Page Mill Rd., P.O. Box 10395, Palo Alto, CA) series high brightness LEDs are officially rated at 100 ma maximum pulse current. We have tested them, however, for reliable, sustained 10 Hz operation at pulse currents of up to 1.4 amps at 1–5 microsecond pulse widths. It is necessary do your own testing with the high intensity LEDs that are anticipated to be used in your system. There is also an analog of Moore's law, which has held true for 30 years, that predicts that the luminous output available from state of the art LEDs will double every 18 to 24 months. It definitely pays off to do a literature search prior to designing your LED array system in order to use the latest parts.

7.5 IMAGING SYSTEM SYNCHRONIZATION

For low-cost imaging systems, the imager most likely to be used would be a standard commercial-grade RS170 output monochrome camera. These have 30-Hertz frame rates with each frame consisting of two interlaced sets of images. Since low cost cameras free run using an internal crystal for its clock, the image acquisition, drop ejection and image strobe must be synchronized to this signal. If an external vertical synch or frame synchronization pulse is not available from the camera, a LM1881N integrated circuit chip can extract one. This 60 Hz (actually 59.997 Hz) synch pulse can be used to trigger with a time delay the firing of the strobe or the ejection of a drop. Digital dividers can take this 60-Hz frame rate and use it to trigger drop ejection and image acquisition on whatever appropriate integer submultiples of this basic rate are optimal for the application.

For instance, if real time image processing is to be done on the image, then it may not be possible to operate at the full frame rate. The fractional charge search experiment at SLAC, which imaged falling drops using a computer interfaced RS170 camera ran at 10 Hz due to the requirement that the image have centroids extracted and paired with images from previous frames in real time. Time delays can be placed inline between the rate-divided pulses and the actual trigger inputs for initiation of drop ejection or strobe operation. Using these delays, the different stages of droplet formation can be studied in detail. Due to the fact that frame interlacing effectively blanks out half of the camera operating time, the use of delays lasting longer than 16 milliseconds should be done on the input to drop ejection rather than on strobe triggering. Otherwise, the strobe may trigger into the next frame. Conversely a blank frame may be due to the strobe being triggered on the wrong interlaced half of the image.

A low-cost circuit that implements these functions is shown in Figure 7.4. The RS170 video signal from the camera is taken off in parallel from the frame grabber input and is fed into the LM1881N video synch extractor. The odd and even field output, which runs at the frame rate (one half the interlaced field rate), is fed into a divide by N counter (the CD4017). A rotary switch then selects a video synchronized clock rate for the imaging and drop ejection that is an integer divisor of the 30 Hz frame rate. The electrical trigger pulses to the strobe, frame grabber and drop ejector are fed through independently settable variable time delays implemented by 74LS123 monostables. Finally, the outputs are buffered by 74F04 inverters.

No frame grabber is needed if one is willing to restrict the imaging update rate to 30 Hz or greater. In practice, one may wish to operate at slower rates to study the restart transients of the fluid drop ejector-fluid combination. Also, some fluid drop ejector-fluid combinations have maximum and minimum ejection rate that can only be found by empirical testing. At rates lower than 30 Hz, the transient flashing of an image onto a dark monitor screen makes it difficult to study the ejected droplet of fluid jet. The ability to trigger on and freeze anomalous drop ejection events is another advantage to imaging with a frame grabber.

The actual time delays between the illumination strobes and the drop ejection pulses produced by the monostables can be measured with an external oscilloscope.

A far more convenient though more expensive way to implement this function is to use programmable pulse generators such as the Wavetek 859 in which the delays between a trigger pulse and an output pulse can be entered on a control keypad or by computer control via an IEEE488 control bus (Figure 7.5).

Background illumination for machine vision imaging where threshold detection and precision centroiding determine velocities requires that the background be uniform to avoid measurement artifacts. Attempting to image drops as dark shadows with the strobe source directly behind the drops ejection path without using a ground glass diffusing screen can yield drop images but also can include artifactual bright and dark regions in the background.

For a uniform background without optical background artifacts, a diffuser, such as a ground glass screen, is needed. This takes out the gross structure caused by non-uniformities in the distribution of the light source but can still leave gradual center-to-edge illumination nonuniformities. In addition, optical vignetting caused by edge obstructions to the light cones, such as when one views through a channel, can cause asymmetric center-to-edge illumination gradients. These illumination gradients can be eliminated by the crude but effective method of selectively blocking the incoming light to the ground glass screen with opaque objects, such as electrical tape and flexible wires placed at different distances between the illumination source and the ground glass screen. An LED array can have the individual LEDs, differentially driven, physically redirected or masked off to optimize the illumination distribution.

To be usable for studying the process of microdrop ejection, a machine vision image display system must be able on receipt of a timing signal to acquire and hold an image on the screen for an indefinite period of time until the next event trigger and be capable of selectively expanding the contrast around a user-defined background brightness level.

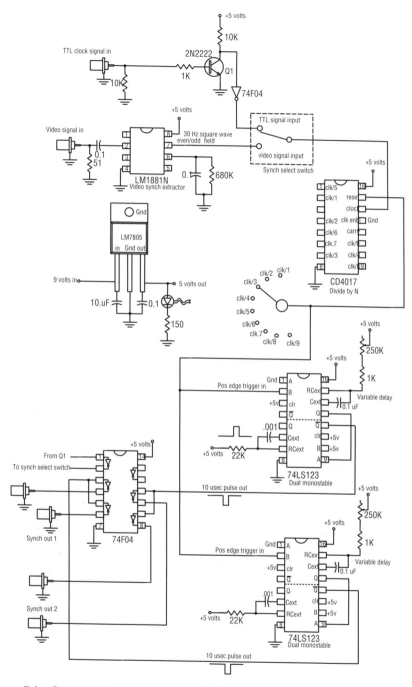

Figure 7.4   Synchronization pulse generator for facilitating the viewing of ejected fluid jets and microdrops. The input is an RS170 video signal from the monitor camera. The output are 10-microsecond-long TTL level pulses with user settable delays for triggering the drop ejector pulse generator and the illumination strobe.

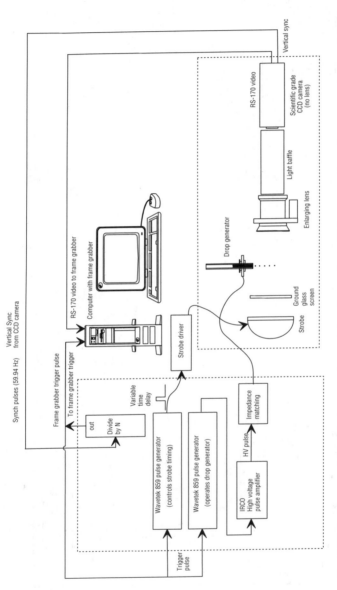

Figure 7.5 Functional block diagram of microdrop generation and imaging system utilizing research-grade, commercially available components. The approximate cost of such a system is in the $20,000 to $30,000 range.

For operation at low repetition rates, the imaging system holds the frame on the screen until the next frame update. This is important since the maximum data rate in experimental scenarios is likely to be less than the frequency needed to produce a subjectively flickerless image. In addition, there are some reasons for going to very low rates, such as to study first-drop phenomenon in fluids that may clog apertures due to evaporation. Frame grabbers can be either interface boards plugged into a computer bus and operated by custom software; or can take the form of a self-contained externally triggered video frame memory unit such as the Colorado Video 441S Frame Store (Colorado Video, Inc., 6205 Lookout Road, Suite G, Boulder, CO).

There must be a contrast expansion and overall image brightness adjustment that is capable of very wide-range operations. Droplet shadows of very small diameter drops against a bright background are often only slightly darker than the background. This may make the drops and the ejection jet nearly imperceptible to the eye. The background, however, can be made extremely uniform in brightness. If the contrast level of the display can be expanded around the intensity level of the drop shadow, the nearly imperceptible microdrops can be made very easy to see. This typically requires a much higher level of brightness and contrast control than is usually available from commercial video monitors and must be built into the display software. The drop diameter threshold where microdrops become difficult to see using background shadowing without greatly enhanced contrast is in the 5- to 10-micron diameter range depending to a large extent upon the imaging system magnification.

7.6 OPTICS

Magnification is a critical factor in determining what can be imaged. There are minimum magnifications that are needed to image drops of a given size. In practice with magnification defined as the enlargement of the real space image to the projected image at the imaging device, it has been found that magnifications between 1 and 4 were sufficient for characterizing the behavior of ejected microdrops.

In experiments, the focal length of the lenses used to image drops has been from 80 to 200 mm. In principle, any focal length lens can give any magnification by appropriately setting the lens to object distance. There are some practical constraints of which lenses can be used in a practical system:

Most microdrop systems operate in a convection shielded chamber or have droplet ejectors positioned by robotic motion stages, which set a constraint on how close the lens can approach the droplet image plane. Some experiments may have additional physical devices that restrict how close the lens can be to the ejected drops. This sets a minimum lens-to-drop image plane distance. If a high magnification is desired, this in turn sets a minimum focal length for the lens.

The F ratio (focal length divided by the lens diameter) affects the image brightness. For instance, using a lens with an F2 focal ratio, as opposed to F5.6, would increase image brightness by a factor of 8, reducing proportionately the pulse width required to produce an image. This either increases the clarity of the images of rapidly moving drops in the high-speed regime or reduces the required number of LEDs in the illumination array.

There are multiple downsides, though, to increasing the speed of the lens. The first is that, for a given lens design, as the F ratio of the lens decreases, the off axis aberrations increase. This can result in inaccurately imaged drops when the drops are off center of the optical axis. These image aberrations can make these drops appear out of focus (curvature of field), nonsymmetrical (astigmatism and coma), with nonlinear correlations to real space displacements (distortion), or with false color (chromic aberration). If quantitative centroid extraction is required, a distortion-free image is required if the center of mass of the drop is to be accurately inferred. If only visual inspection is required, the optical field quality may be appreciably relaxed. Another relevant issue is that high optical quality fast lenses are available only at very high cost (thousands of dollars).

Conventional camera lenses are corrected to focus light to a point for objects located at infinity. For objects located at finite distances, a lens corrected for objects at infinity will exhibit spherical aberration. However, photographic enlarging lenses are made for transferring an image from a finite distance to another finite distance. This allows the image to have less spherical aberration for droplet imaging applications than one corrected for imaging an object at infinity. This is a matter of degree rather than an absolute difference. Even enlarging lenses are only corrected for one image-to-object distance. In practice, one minimizes the spherical aberration problem at different distances by using slow F ratios. If the iris is closed enough to produce a large enough F ratio, the spherical aberration and nearly all other off axis aberrations can be eliminated. The trade off is that image brightness is reduced requiring more powerful illumination that in the upper limit can produce convection currents disrupting the motion of the drops in addition to heating the drop generator fluid that can alter its rheological properties. If the aperture is reduced further, resolution will degrade even if the illumination is increased to compensate.

There is, in practice, an optimal iris aperture opening for real world (i.e., imperfectly aberration corrected) lenses. If the aperture and illumination level are too small, diffraction-limited imaging suffers. If they are too wide, the lens aberrations start to become objectionable. Chromic aberration can be effectively eliminated by using a monochromatic filter over a gas strobe source or by using pulsed LED illumination. There is quite a noticeable increase in image sharpness when a monochromatic source is used.

Since enlarging lenses used in their originally intended application transfer a stationary image to a stationary focal region, there is no real requirement for fast F ratios. These lenses typically are available in F ratios of about F5. For the primary goal of imaging high-speed phenomenon where moderate amounts of off axis aberrations can be tolerated, a low F-ratio commercial photographic lens in the range of F1.2 can be used. Going from F5 to F1.2 reduces the required number of LEDs by a factor of 17 or allows an equivalent reduction of the pulse width.

7.7 CAMERAS

For imaging microdrops, monochrome, solid-state cameras are the instrument of choice. Color cameras lose about a factor of ten in sensitivity and often use imaging chip multiplexing schemes that make it difficult to interpret the image on a pixel-by-pixel basis, which is needed for drop centroiding. There are few microdrop applications where the addition of color adds any information to the image.

In general, the larger the imaging chip for a given imager resolution, the greater the sensitivity. Currently, CCD-based imaging chips are more sensitive than comparably sized CMOS camera chips, but this is subject to change as technology develops. Standard RS170 output, black-and-white video cameras give very acceptable drop images. There is no need to go to expensive scientific-grade, solid-state cameras unless precise centroiding over very large areas of real space is required. Experiments done at Stanford Linear Accelerator Center tracking the motion of falling fluid drops to search for particles with fractional electric charge used commercial-grade Cohu 748 × 484 pixel RS170 output monochrome cameras (Cohu, Inc., 12367 Crosthwaite Circle, Poway, CA) and analog frame grabbers. Using these commercial-grade cameras, sophisticated image processing software running in real time on 200 MHz Pentium® II (Intel Inc., 2200 Mission College Blvd., Santa Clara, CA) class personal computers were able to find and centroid multiple drop images at 10 Hz to better than 1/30 pixel accuracy.

Low light sensitivity is more important than the raw number of pixels in obtaining the detailed images of microdrops in high-speed fluid jets. This is because high-speed events require short-duration stroboscopic illumination. Most stroboscopic illumination sources have a proportional relationship between optical energy per pulse and time duration of the pulse. The less optical flux needed for an image to be detectable over the camera background noise, the faster and shorter the illumination pulse can be, which results in sharper details of the rapidly moving fluid jet.

The ability to disable any automatic gain control or image brightness controls on board the camera is the one essential feature of a solid state camera that is used for drop imaging. The optimal background illumination levels for microdrop imaging are such that the camera will try to fight attempts to optimize the average intensity to levels that are best for viewing small, low-contrast drops. Automatic gain and background adjustments also make the image brightness unstable. This renders machine vision analysis of the images to extract drop centroids difficult.

To obtain the proper magnification for most applications, the camera lens-to-microdrop focal plane distance is sufficiently large that it is well out of the practical distance at which one can focus the optics if the lens is attached conventionally to the camera. The camera-to-lens distances typically used at SLAC are in the range of 20 cm to over 1 meter. Depending upon the optimal magnification for viewing different drop phenomenon, this distance is subject to large changes. For instance, for viewing details of the ejection jet, a high magnification necessitating a large lens-to-camera distance is required. In contrast, for viewing the ejected drops to determine rates of evaporation, a wide field of view is needed, which could reduce the camera-to-lens distance to 10 or 20 cm. In research settings the camera and lens are often mounted on an optical rail to obtain the needed flexibility in relative positioning. An additional component that is needed is a light, tight shield running from the camera to the lens. Otherwise, diffuse light hitting the imaging chip will greatly reduce contrast and an unstable average background illumination level will make machine vision centroiding difficult. Light-tight tubes were constructed of black cardboard and attached to the lens and camera with black electrical tape in order to have an inexpensive source of easily mounted light baffles. These cardboard tubes can be made to telescope inside each other to make a variable length baffle.

If one is using imaging in an application where a fixed magnification is usable and the required magnification allows a relatively short camera-to-lens distance, the lens can be coupled to the camera with a series of C-mount extender tubes. This single piece setup is far easier to mechanically position than a multiple element setup on an optical rail. It also facilitates a high-contrast image by reducing the amount of scattered light onto the camera imaging element.

When viewing backlit drops with enhanced image contrast, dust particles on the CCD sensing element, which are invisible when used to view objects under normal noncontrast enhanced conditions, will be highly visible. This is both visually annoying as some of the fine particulates can look like drop images, but more importantly, a large number of dark spots on the image can impede the operation of machine vision drop finding programs. The obvious solution is to clean the CCD and then to seal off the opening so that dust contamination can be kept out.

Cleaning the CCD chip must be done with caution. A scratch on the surface of the CCD sensor chip by a particle of dust in the cleaning cloth can permanently ruin the camera. It is surprisingly hard to clean the surface of the chip without redepositing dust, which happens if an attempt is made to clean with compressed air, or leaving streaks, which may happen if simple solvent moistened wipes are used. We have found one method that consistently works:

- Make a wipe from a cotton tip swab that has optical cleaning paper folded over it in such a way that the surface at the end is flat and the width of the CCD.
- Moisten the wipe with a lens cleaning solution, such as AOSafety Superclear® (Aearo Company, 5457 West 79th Street, Indianapolis, IN). The wipe should be moistened with just enough fluid to allow it to retain dust by fluid-surface tension but not be excessively wet or fluid streaks and evaporative rings will be left.
- The CCD should be staring into an illuminated area with the intensity set such that the pixels are not saturated. The viewing system should be on and set at high contrast so the dust particles on the CCD surface are easily visible. This is the only way to know if the CCD is clean. Physical examination of the surface of the CCD is inadequate.
- Make one continuous wipe across the CCD, starting at one side of the CCD chip, and then check the image to see if the dust was removed. If linear streaks in the direction of the wipe appear, it is usually a sign of too much cleaning fluid on the swab. The swab should only be moist enough to attract and retain dust particles. Insufficient cleaning fluid usually results in dust particles being relocated, not removed. Usually it takes several wipes before all the dust is removed and no fluid streaks are left.
- To maintain a clean surface, cap the open end of the camera with an optical window. A removable optical window can be improvised from easily available components by purchasing a C-mount thread extender and gluing onto it a cut section of an antireflection-coated, clear photographic filter window. These are clear glass protective covers that screw into the filter threads of SLR camera lenses. They are intended to act as a sacrificial surface to be scratched instead of the lens itself.

7.8 STROBOSCOPIC ILLUMINATION SOURCES

Viewing the details of the fluid jetting and microdrop formation process requires some method of acquiring images of the drops over very short time frames. Stroboscopic illumination, in which very short duration, high-peak intensity light sources are used to illuminate the microdrops, is the most practical method of acquiring these images. The alternative approach of using continuous illumination and shuttering the camera is impractical due to the extreme level of illumination required for microsecond-long exposures to produce images on current generation electronic cameras and photographic film.

One can readily calculate why such short pulse rates are needed. The ejection speed of the fluid jet out of microdrop ejectors ranges from 1 to 10 meters per second. For instance, a fluid jet ejected with a 1-meter-per-second initial velocity illuminated with a relatively long 100-microsecond pulse, would smear out the image of the jet over 100 microns of travel. Since the sizes of the drops of general interest to experimenters range from 1- to 100-microns in diameter, this 100 microsecond long pulse would render invisible the fine structure of the ejection jet and the formation process of the microdrops. If one wishes to study the process of droplet formation, a very short, intense illumination source capable of providing enough integrated output to form an image over a time duration of less than 5 microseconds is required. For imaging microdrops traveling at their terminal velocity in air, one can use much longer illumination pulses. The terminal velocity of fluid microdrops in air is on the order of a few mm to a few cm per second. Also, the principle reason for imaging microdrops outside of the jet formation region is to verify that they are still being reliably generated and to track their motion. A small amount of distortion in their direction of travel does not interfere with these tasks.

The specifics of the illumination system depend upon the sensitivity of the imaging device. In order to generate the short pulses needed to image details of the high-speed fluid jet, the low luminosity of the LEDs available then, and the relative insensitivity of first generation CCD cameras (early 1990s), required peak optical intensities that only spark discharge strobes and pulsed lasers could supply. Lasers ended up being poor illumination sources for machine vision applications due to the coherent monochromatic light producing self-interference patterns that produced shifting pseudo periodic bright and dark spots throughout the image.

The pulse durations from commercial xenon strobe lamps were dependent upon the flash intensity. At higher flash energies an increased amount of heated gas must cool in order for the optical emission from the flash lamp to cease. At the intensity needed for drop imaging, the measured pulse width was about 1 microsecond.

LEDs are not subject to the gas strobe's wave shape and pulse duration issues resulting from thermal time constant limitation on the minimum pulse widths that they can generate. The intrinsic band width of LEDs allow modulations of the light intensity by the applied current in the gigaHertz range. LEDs are constrained in their usable pulse widths by their relatively much lower peak optical output. When using first generation CCD cameras, in order to generate enough integrated illumination to supercede background noise using pulsed LED arrays, a pulse length of at least 50 microseconds is required. This is short enough for drop centroiding and motion analysis, but not for studying the process of drop formation from a fluid jet. Using recently manufactured high brightness LEDs and state-of-the-art (2001), high-sensitivity CCD cameras (0.01 lux rating), only 3- to 5-microsecond pulse lengths were required from the LED array. The fine structure of the ejection jet was readily visible at these illumination pulse lengths.

Low-light CCD cameras are currently available for less than $500. Triggerable, research-grade Xenon strobe illuminators cost about $2000. An LED illumination array and driver can be assembled for less than $100 in parts. While in 1992, when we began our research, the lack of sensitivity of the available cameras and the low efficiency of the then-available LEDs forced us to use gas strobe illumination sources. The output levels of currently available LEDs and the sensitivity of current low-light, solid-state cameras, which now allows microsecond illumination pulses from LED arrays to image fluid jets, calls into question whether gas discharge strobes are now obsolete for this application.

There are other reasons favoring the use of pulsed LED arrays over gas strobes. The flashlamps in gas strobe systems, in addition to costing in the hundreds of dollars each, have a limited operating life. Bulb life under continuous laboratory operation can be as short as a month. In contrast, an LED array and driver can be assembled out of easily available components for less than $100 in parts. When operated within safe current limits, LED arrays have essentially indefinite operating lives. LED arrays also have the advantage in machine vision systems of maintaining uniformity of illumination over time.

Because gas strobes operate using a high voltage arc within an enclosed glass envelope, each discharge vaporizes a portion of the electrode and deposits the vaporized metal on the interior of the bulb. The strobe bulb glass envelope slowly darkens with time, and the position of the arc within the bulb can shift from flash to flash. For machine vision applications where the drops are located by subtracting out the background, this flash-to-flash variation in intensity and distribution of light can cause erratic operation of the object detection software. LED arrays also give sharper images since they are predominantly monochromatic sources, unlike gas strobes. Multielement lenses, in practice, only give exact chromatic correction for two wavelengths. We have found that with the same lens, LED illuminated drops produced sharper images. If the strobe, however, is sufficiently bright, a narrow band color filter can be used to sharpen the image. However, higher strobe energy in a given bulb tends to proportionally lengthen the duration of the flash, which partially negates the advantage of using a gas strobe for microdrop illumination (Figure 7.6).

Another issue relating to the illumination requirements is that the type of camera used makes a large difference in what illumination levels are required. The difference in sensitivities between the high- and low-cost, solid-state cameras can literally be orders of magnitude. Between the low-cost, sub-$50 CMOS cameras on a chip and the scientific-grade, multi-thousand-dollar CCD cameras we have used, was a difference of a factor of 100 in specified light sensitivities. This is reflected in the differences in the illumination pulse widths required when using these cameras with LED arrays. One step up from the one-chip CMOS video cameras is the multichip, single-board, commercial CCD cameras selling from $100 to $200. The low-cost CCD camera pictured in Figure 7.10 with the backlit screen in close proximity (4 cm)

Figure 7.6   LED array strobe illuminator and drive electronics.

to the drop ejector, required an illumination pulse of 20 microseconds when illuminated by a 24-element LED array. A Cohu scientific grade CCD camera (6310 series)
with the same LED array, but with a backlighting screen located 10-cm away,
required a pulse width of 75 microseconds. An industrial low-light-rated CCD
camera, Watec WAT-902C (Watec America Corporation, 3155 E. Patrick Lane, Las
Vegas, NV) required only a 5-microsecond pulse width. You definitely get what you
pay for in using higher quality cameras, but for some applications that only require
verification of drop ejection or for tracking the motion of drops after reaching
terminal velocity, there may be no need to use the more costly technology.

The circuit in Figure 7.7, shown assembled with a 24-element LED array in
Figure 7.6 is a minimum-cost illumination pulse generator and LED driver. It is
designed to generate optical pulses from 1- to 150-microseconds in length. The
circuit is crude and pays for its simplicity with slightly nonrectangular pulses. As a
consequence, the amplitude for very short pulses can be a function of the pulse
width. Its saving graces are that it is simple, cheap to construct, and it works!

The circuit takes a TTL timing pulse as an input and uses the positive edge as
a reference to trigger the 74LS123 monostable to produce a pulse with a width set
by the 100K potentiometer. The output pulse is sent to the IRF3205 power MOSFET.
This transistor is a remarkable part that has a peak continuous rating of 110 amps
and a peak pulse current of 390 amps. The LEDs are being driven at a peak current
(dependent on pulse width) of 200 milliamps each for a total of 4.8 amps for the
array. When operating an LED in high-current pulse mode, the normal voltage drop
specified by the manufacturer's data sheets for continuous operation seriously underestimates the voltage drop across the LED. The series resistor must have its value
specified for the pulse current levels at which they are intended to operate.

The pulse width is set to a specified range with a hardware-limited maximum
pulse length. This was to prevent destruction of the LED array by operating it at

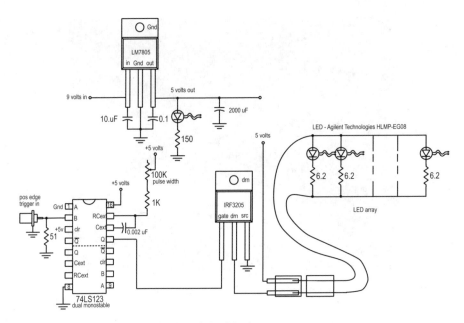

Figure 7.7   LED array drive circuit optimized for low cost.

too high a time-averaged power. An alternate method of setting the pulse width is by using the TTL trigger's pulse width as a gate to determine length of the optical pulse. Its advantage is the pulse width can be set under computer control for machine vision applications. Also, if it is necessary to use multiple modules synchronously to create a massively parallel illumination array, using the trigger pulse width as a gate to define the pulse width allows the illumination array to be synchronously varied in pulse width from a single-control source.

There are applications where the simple circuit in Figure 7.7 is inadequate, specifically if very short pulse widths less than a microsecond are needed. The circuit in Figure 7.8 was designed by physicist Valerie Halyo of SLAC to be used as a modular drive element in a 44-drive channel, 1056 LED illumination array. The circuit can produce a rapid sequence of submicrosecond-long pulses at 20- to 100-microsecond intervals to illuminate rapidly moving, large, high-density fluid drops to extract trajectory information from multiple exposure tracks on a single CCD frame. In a way this is an updated version of the photographic film and a strobe wheel method Hopper and Laby[1] used in the 1940s to track the motion of fluid drops. The 1056 LED array system was built to track drop velocities that were a few orders of magnitude higher than the millimeters-per-second drop speeds that the strobe wheel system was capable of handling.

Each board drove 24 LEDs. The rise and fall times are on the order of a few tens of nanoseconds depending upon the load. This is a high-speed circuit in which interconnect parasitics can seriously degrade performance. This circuit must be built on a properly laid out, printed circuit board.

Unlike the simple edge-triggered circuit in Figure 7.7, this high-speed LED driver uses the input TTL level pulse as a gate and switches on the LED array for

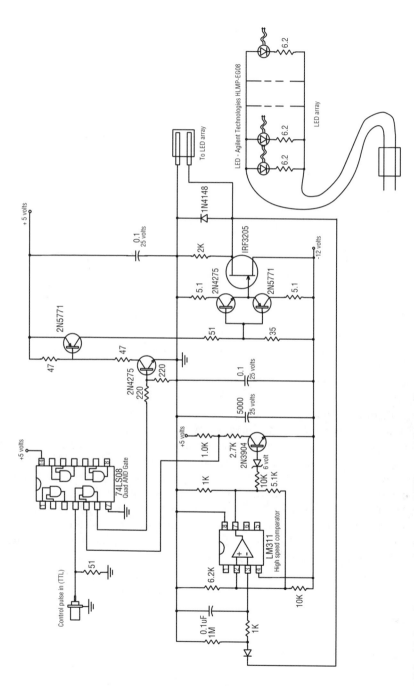

Figure 7.8   High-performance stroboscopic LED array drive circuit.

an interval equal to the time high of the control pulse. A pulse time integration and comparator outputting to an AND gate circuit limits the total pulse length to 100 microseconds, protecting the LEDs from destruction by excessive pulse length. The output drive transistor is the same as that used in the simpler circuit, but the drive to the MOSFET gate is implemented by a fast NPN PNP transistor stack. This circuit, when constructed on a point-to-point wired prototyping board, exhibited some ringing that disappeared when built on a printed circuit board laid out using a ground plane and minimum trace lengths between components.

7.9 LOW-COST MICRODROP IMAGING HARDWARE

Visual observation of back-illuminated drops is sufficient to verify ejectability and to perform alignment operations. However, a detailed quantitative study of drop ejection dynamics requires some type of imaging where the drop images can be recorded and studied in detail. This does not necessarily require the use of electronic imaging.

In order to track the trajectories of falling drops to measure their electric charges, Hopper and Laby[1], in 1941, conducted a drop-imaging experiment to determine the value of the electronic charge. They used photographic film to obtain trajectory information of 3- to 10-micron diameter fluid drops falling in air. The drops were illuminated by a 20-amp carbon arc lamp strobed by a rotating disc that had a transparent slit with a 6-degree angular width. A volume of water was placed between the arc lamp and the beam collimating optics, in order to absorb the heat produced by the carbon arc. A series of lenses and baffles were used to produce a well-defined beam in order to image the drops as bright objects against a dark background. The multiple exposures of the same drop on the film were then inspected with a metrology microscope to extract their velocities. This system produced effective strobe pulses with 1/1500 of a second widths at a rate of 25 Hz. A conventional shutter was used to select the total exposure time and number of strobed images. The exposure time for their drop-tracking experiment was two seconds.

Highly accurate quantitative data was obtained with this photographic plate-strobed illumination source method, but it has a major problem in studying drop ejection: the time delay between making a change in the system and being able to see the results. Modern electronic imaging systems utilizing video cameras and frame grabbers operating synchronously with the drop ejection drive pulses allow instantaneous feedback on the effects of changes in the drop ejector's operating conditions on the drops ejected. Computer interfaced cameras also allow for real-time quantitative determination of drop ejection characteristics. The recent availability of inexpensive, solid-state video cameras, frame grabbers, and high-brightness LEDs has made electronic microdrop imaging systems relatively cheap to construct (Figures 7.9 and Figure 7.10).

7.9.1 Low-Cost Cameras

One simple implementation of electronic imaging utilizes a commercial CCD monochrome camera selected for good low-light response and its ability to shut

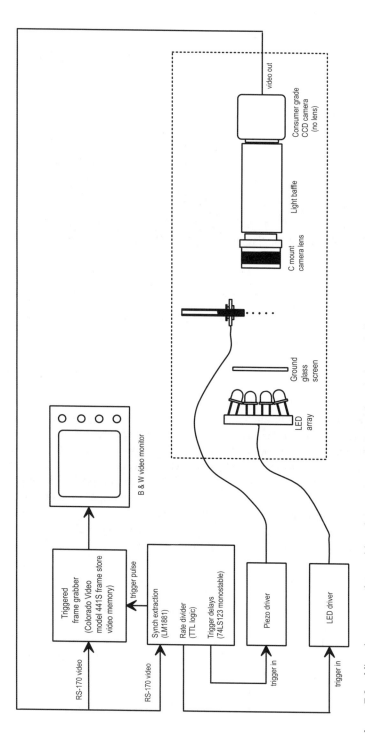

Figure 7.9 Microdrop generation and imaging system constructed from in-house built components and consumer grade parts. The cost of a microdrop generation and imaging system constructed in this manner, neglecting assembly costs, is in the $2,000 to $3,000 range.

Figure 7.10   A minimum cost implementation of the block diagrammed system in Figure 7.9, using a combination of in-house constructed and consumer grade electronics and optical components. The camera is a sub-$100, commercial, monochrome security camera. The lens used is a 50-mm, C-mount, commercial-grade lens. The light tight tube coupling the camera to the lens was constructed from black construction paper and electrical tape. Clear images of the drop ejection process could be obtained with this simple low cost set up.

down automatic gain controls. The better the low-light response of the camera, the shorter the illumination pulses that are required to produce an image (Figure 7.11). Shortening the illumination strobe flashes increases the amount of detail seen in images of rapidly changing phenomenon, such as the break up of high-speed fluid jets into microdroplets. It is necessary to be able to disable the automatic gain features of most cameras, since the optimal average illumination levels that are needed to view microdrops are usually not at the point where most cameras will try to set the average background value of its image. As of 2002, an acceptable camera can be purchased for around $50 to $100. A high-sensitivity, low-light camera costs about $500 to $700. Some money can be saved by purchasing an camera without an integral lens. A practical system will require an external lens mounted from 10 to 50 centimeters away from the camera focal plane.

7.9.2   Low Cost Optics

The lens in a low-cost system can be a conventional C-mount, commercial-grade video camera lens. In principle, high-grade photographic enlarging lenses are better, because they are designed to transfer an image from a close focal plane to another close focal plane with magnification factors approximating the magnifications used for drop viewing. However, the cost of good enlarging lenses is in the hundreds of dollars. In contrast, a new commercial-grade, C-mount camera lens can be purchased for about $50. Optimal focal lengths are in the 50 to 135 mm range. The focal length is constrained in the long direction by the cost of the lens for a given F ratio and

Figure 7.11   A high performance implementation of the system block diagrammed in Figure 7.9 using research grade components. A premium-quality, low-light camera is used to allow reducing the LED strobe flash duration to the microsecond range for increased clarity of imaging the fast-moving fluid jet. To maximize contrast, solid threaded coupling tubes are used to provide a closed-light path between the lens and the camera. An enlarging lens was used as the image transfer optics to optimize imaging near focal objects. To minimize contrast loss due to scattered background lighting, a filter is attached to the imaging lens with a spectral band pass matching that of the emission spectrum of the LED strobe. The synch extraction, timing, and LED drive electronics were built into a single standard, 19-inch, rack-mountable chassis.

the increased length of the optical rail. However, if the focal length is too short and the drops are being ejected inside of a convection-controlled chamber, the lens may not be able to reach the focal plane of the drops at a desired magnification. Also for a given F-number-long focal length, a lens, in general, will image with less distortion and fewer optical aberrations. The lens used in the system pictured is a 50-mm focal length F1.4. The focal lengths of the lenses used in our scientific experiments have been in the 135-mm to 200-mm range with 135-mm being the most commonly used lens focal length.

REFERENCE

1.   V.D. Hopper and T.H. Laby, The electronic charge, *Roy. Soc. of London, Proceedings, A*, 178, pp. 243–272, 1941.

CHAPTER 8

Drop Ejector Drive Electronics

The two basic types of electronic circuits used to drive drop-on-demand piezo-electric elements are gated voltage sources and high voltage linear amplifiers.

Piezoelectric elements that are used to actuate drop ejectors can be highly reactive low impedance loads. In addition, the pulse widths that the drive electronics must furnish are on the order of 1 microsecond or less at voltages of up to 1000 volts. Cost effective, low-duty cycle, high current, high voltage pulses at MHz bandwidths require specialized circuits.

The voltages needed to drive drop ejectors vary by orders of magnitude depending upon the drive geometry of the drop ejector and the fluid that one is trying to eject.

For instance, one microdrop ejector that ejected water using amplitudes at around 15 volts, required drive pulses on the order of 800 volts in amplitude to eject high viscosity 60–100 cS oils.

The geometry and size of the piezoelectric element also affect the drive voltage needed. Decreasing the area of the piezoelectric element that contracts on the fluid reservoir walls roughly increases linearly the amplitude of the drive pulse needed.

The amount of fluid in a drop ejector also affects the drive voltage needed. Tubular piezoelectrically driven drop ejectors that use circumferential piezoelectric drive elements have their required ejection amplitude increase dramatically when the fluid level falls below the point at which the piezoelectric element is mounted.

Typically pulse amplitudes of 50 volts or less are needed to drive drop ejectors with low viscosity fluids and optimized geometries and fill levels. High viscosity fluids or fluids that are so costly that the drop ejector cannot be filled to the level of the drive element can increase the required drive level to hundreds of volts.

The drive pulse amplitude can be limited by the breakdown of the fluid by cavitation, which produces gas bubbles that accumulate and suppress drop ejection as well as change the required amplitude as they build up. Glass body ejectors also can be shattered by too high an ejection pulse amplitude. At the Microdrop Particle

Search group, researchers have blown the tips off of their drop ejectors when attempting to clear clogged apertures by brute force overdriving of the drop ejector. Destruction of drop ejectors by overdriving required drive pulse amplitudes in excess of 1000 volts.

8.1 GATED VOLTAGE SOURCE PULSERS

The simplest type of circuit that will perform this task can be configured as a regulated DC power source gated by a high voltage high current transistor stage. IRCO (Instrument Research Company, 7513 Connelley Drive, Suite C, Hanover, MD) manufactures commercial gated voltage source pulse generators. For low-cost implementations, the circuit in Figure 8.1 will drive the piezoelectric elements in drop generators that do not require more than a few hundred volts amplitude.

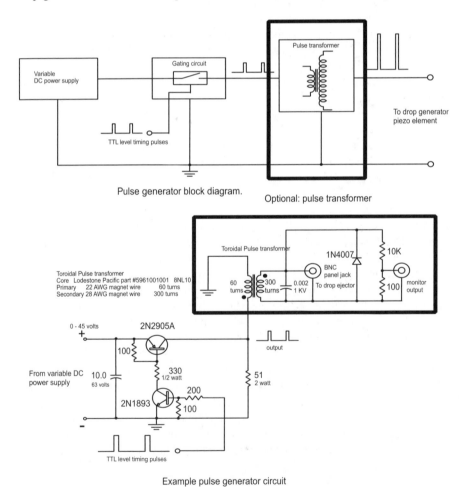

Figure 8.1 Gated voltage source pulse generator schematic.

The circuit is a DC gate using a series PNP transistor in a common emitter configuration driven by an NPN transistor. The pulse width is set by the width of the external input pulse. The DC power source can be any variable DC supply ranging from bench top power supplies to the output of a battery driven variable output IC regulator, such as the LM317. The power rating of the DC supply should be sufficient so that the voltage does not sag at the pulse repetition rate at which one anticipates to use the drive unit.

The output of the transistor switch is fed to a pulse transformer to increase the amplitude to allow driving drop ejectors requiring more than the directly switched 45 volts. The capacitor and diode on the output of the pulse transformer are optional parts used for optimizing the output pulse shape for satellite free ejection for specific fluids.

A complete handheld battery powered unit for driving drop ejectors can be constructed using this pulse output circuit as the final output stage and common LM555 monostable chips as the pulse width, and pulse repetition setting elements (Figure 8.2). The unit also can be operated as an externally triggered pulse generator and in a continuous operating free running mode. For low viscosity fluids with favorable ejection characteristics, this unit is adequate. For fluids that require more precise control over the drive waveform in order to produce stable monodisperse drops, a digital pulse or general waveform generator driving a linear pulse power amplifier is the recommended optimal electrical excitation source for the piezoelectric element.

8.2 LINEAR PULSE POWER AMPLIFIERS

Fluids that have unfavorable rheological characteristics for stable ejection usually require more precise control over the drive to the piezoelectric elements. Alterations in the rise and fall times of the pulses, as well as the capability to use pulse chaining and complex pulse shaping, can be required to produce satellite free ejection of drops or to allow real time modulation of drop diameter. Using linear power amplifiers to drive the piezoelectric elements, as opposed to gated voltage sources, provides this added control. Another advantage to using a feedback-based driver for a piezoelectrically-driven drop generator is that it tends to suppress mechanical ringing of the drop generator structure.

Linear feedback amplifiers made from high voltage operational amplifiers, such as those designed and sold by Apex Microtechnology Corporation, 5980 N. Shannon Road, Tucson, AZ, that have slew rates around 1000 volts per microsecond with voltage ratings up to 450 volts, can be used to drive the drop generator piezoelectric elements directly.

A very robust higher power output pulse power linear amplifier, which has sufficient bandwidth and drive capability for use with piezoelectric elements used in relatively large drop-on-demand drop ejectors, was designed at SLAC by engineer Paul Stiles. The design is an improvement over the more simple circuit in Figure 8.3 in that it has a factor of ten higher bandwidths, improved ability to drive lower impedance loads and a higher peak output voltage.

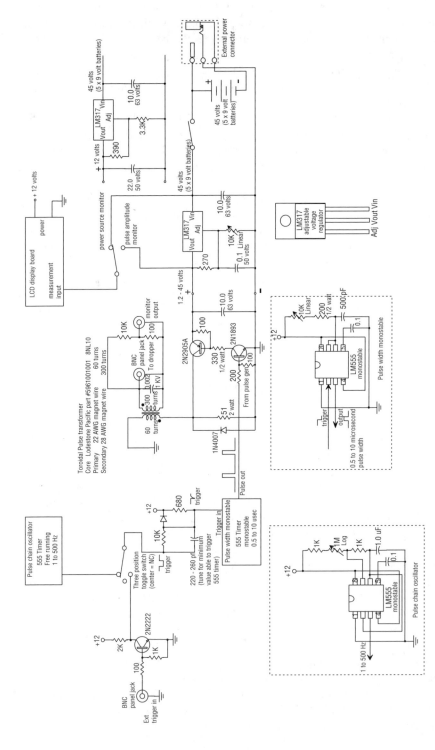

Figure 8.2 Schematic for a low-cost compact self-contained piezoelectric drop ejector drive unit.

   Active voltage feedback is used to stabilize the output amplitude against distortion caused by mechanical ringing of the piezoelectric element and highly reactive loading. The linear amp is stable with output amplitudes up to 500 volts. The design was made possible using reasonably priced components by the constraint that the duty cycle for peak power operation be low, that is less than 1%. Otherwise power dissipation would shut down the amplifier. Drop ejection stability when using fluids with poor rheological characteristics is much better with this type of drive unit than with the gated DC switch type pulse generator. The only practical disadvantage to using this type of drive over a top-of-the-line commercial gated DC switcher is that as the time of this writing there is no commercial source for this type of pulse power linear amplifier (Figures 8.4 and 8.5).

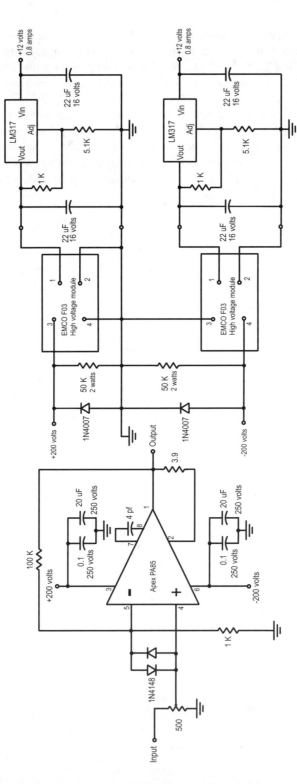

Figure 8.3   Basic linear piezoelectric element drive amplifier. This circuit is built around one of the high voltage, high bandwidth operational amplifiers manufactured by Apex Microtechnology. This circuit has a bipolar output capability which is valuable in implementing fill before firing fluid ejection cycles which requires actively being able to controllably expand and contract the volume of the fluid ejection chamber.

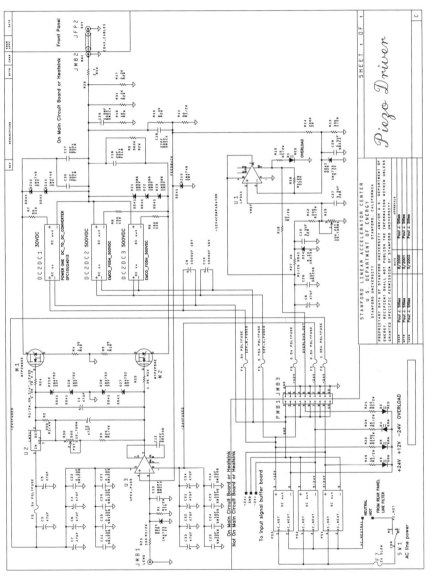

Figure 8.4   Schematic, (diagram 1 of 2) for a linear power amplifier optimized for driving microdrop generator piezoelectric elements.

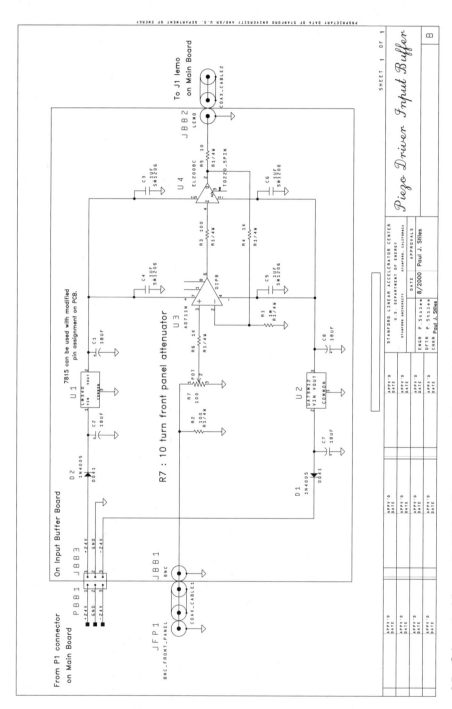

Figure 8.5 Schematic, (diagram 2 of 2) for a linear power amplifier optimized for driving microdrop generator piezoelectric elements.

CHAPTER 9

Fabrication of Ejection Aperture Nozzles

With the exception of the focused acoustic beam technique, all microdrop generation techniques require a part of the drop generator hardware: an ejection aperture nozzle consisting of a low fluid impedance hole that is the same diameter of the microdrops to be generated. Optimally constructed low fluid impedance nozzles in the 10- to 100-micron range are not, strictly speaking, off-the-shelf components, but there are some commercially available parts that can be utilized, and there are a number of ways with varying degrees of cost and complexity to fabricate microdrop ejection nozzles.

9.1 EJECTION NOZZLE REQUIREMENTS

An ideal ejection aperture hole would have a tapered cross section ending up in a short cylindrical hole with an aspect ratio of about one-to-one. The purpose of the conical overall taper is to minimize the fluid impedance without compromising mechanical rigidity if the structure is to be formed on a flat plate (i.e., silicon wafer). Standard silicon wafers used in microfabrication processes are about 500-microns thick. Thinner wafers are available but at a much higher cost and are processed with a much higher chance of breakage.

Compared to a tapered hole, a straight, cylindrical 20-micro diameter hole through a 500-micron thick wafer would have an excessively high fluid impedance requiring a much higher drive impulse to affect ejection as well as to limit refill rate for the fluid near the tip, which reduces the maximum rate for drop ejection. For instance, tapering the hole from 500 microns at the reservoir end to 20 microns at the outside surface would drop the fluid impedance and increase ejection efficiency. A nearly cylindrical section near the exterior is desirable so that capillary refill can be used to drive out any injected air reducing "gulping," which for some excessively violent ejection fill cycles can result in air being taken into the drop ejector near the

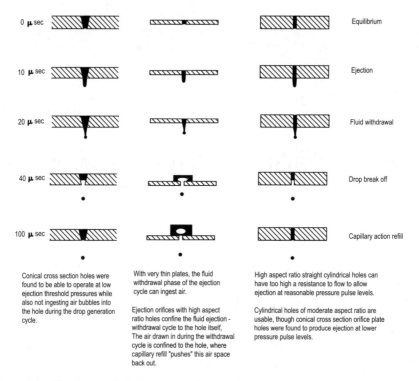

Figure 9.1   Ejection nozzle hole cross section requirements.

aperture, which raises the ejection threshold due to air bubbles absorbing the pressure impulse meant to drive out the fluid. Some commercial printhead designs incorporate an internal step where the fluid under negative pressure can assume a stable position in order to have a dry exterior surface to minimize particulate build up from evaporation, and a stable secondary internal position for the fluid under negative pressure to have a more repeatable ejection cycle (Figure 9.1).

9.2 FABRICATING ACUTE EDGE CONICAL GLASS NOZZLES

One common method of making a reservoir tube with a micron scale ejection hole is to heat the glass tube and pull it while hot in a manner that produces a closed cone with an included angle of about 30–45 degrees. This pointed conical end is then ground back until a hole of the desired diameter is exposed.[1]

There are various ways of obtaining an axially symmetric conical terminated end on a length of glass capillary tubing. Simply heating a length of tubing and pulling by hand usually results in very nonsymmetric conical termination. A better method is to attach the glass tubing to a rotating chuck while heating. One simple implementation is to make an adapter collet to a low-speed electric motor mounted such that the glass tube is vertically oriented. A weight is attached to the glass tube

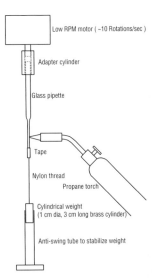

Figure 9.2   Low-cost setup for producing a conical closed end on a glass pipette.

by a thread. If too much heat is used over too large an area, the cone produced is very long and sharp. Too much heating, and a long taper with a glass thread connecting the two halves of the separated glass tube will result. Low, localized heating gave the best results. The torch should be set for a very small flame set at one small spot on the rotating glass tube. With the flame set at the right distance so that the rate of heating is correct will result in an abrupt separation of the glass tube into two halves each with a short closed conical sealed end. This works and is inexpensive to set up, but it depends upon a lot of trial and error hand work to get the glass to separate to produce the right cone angle (Figure 9.2).

A better, more reliable way to obtain a closed end conical tip in a glass pipette is to use a modified conventional biology micropipette electrode puller. This has the advantage of results being far more repeatable than for the case where a manually held hand torch is used. The major disadvantage is the cost. The lowest priced units that have sufficient versatility to be adapted to making microdropper pipettes start at $3000 to $4000. SLAC researchers eventually set up a Sutter Instruments P-30 glass micropipette electrode puller (Sutter Instrument Company, 51 Digital Drive, Novato, CA) to perform the task of sealing off the glass pipettes with the correct termination angle at the tip (Figure 9.3).

The factory supplied heater wire was modified by removing one turn and compressing the remaining two coils of wire together (Figure 9.4). This was done to increase and localize the heating. This is necessary because the usual glass pipettes for which the machine was designed was about three times smaller in diameter than the end of a Pasteur pipette. The thermal processing in the Pipette puller is done in two steps:

- The glass is heated to the point where separation and constriction of the end occurs. The heater setting used on the P-30 for this step is 799. The pull tension setting is 50. The end result of this step, however, is not the desired closed, abrupt cone but a sharp, tapered, open-ended cone (Figure 9.5).

Figure 9.3   Sutter Instruments P-30 glass pipette puller. While not originally designed to produce closed end 30 to 50 degree internal angle terminations a slight modification of the heating elements and a two stage operation consisting of a standard pull followed by melt back step, can repeatably produce properly terminated glass pipettes.

Figure 9.4   Close up of the heating element of the Sutter P-30 pipette puller with a Pasteur pipette in place. The P-30 is not as designed intended for use with this diameter of glass tube nor with the intent of producing an abrupt closed end termination. The heating element shown in the photo is compressed into shorter length coil than what the factory originally wound it in order to concentrate the melting of the glass to a narrower zone.

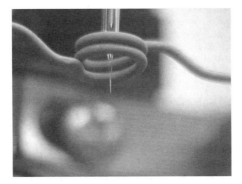

Figure 9.5   The result of the initial heating and pulsing step of the Sutter P-30 produces in the glass tube a very acute sharp edge open ended termination to the pipette which is suitable for acting as a glass hypodermic needed for intercellular electrode work (the original design purpose for the puller) but not as the starting end cap geometry for a microdrop generator nozzle.

- With the pipette in the same location, it is heated in place to melt back the tip and close off the end. The Sutter Instruments P-30 is set for maximum heater current (heater setting 999) for two minutes, timed manually, to do this melt back step.

The duration of this post heating determines the degree of tip melt back and hence the included tip cone angle. The desired included angle is usually between 30 and 45 degrees (Figure 9.6). This closed end is then sanded on fine abrasive paper (600–2000) until a hole of the desired diameter is cut. The tip grinder that we use consists of a MotoTool® (Dremel, 4915 21st Street, Racine, WI) with a 5-cm diameter sanding disc cut from 600 grit carborundum paper. The sanding disc is mounted within the field of view of a low power binocular microscope (Figure 9.7). A small V groove block held the glass tube in a perpendicular alignment to the sanding disc.

One method of determining when to stop grinding is to insert a length of tungsten drift chamber wire the diameter of the desired hole into the tube pressing against

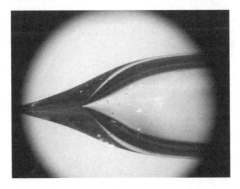

Figure 9.6   Tip of the pipette after the second heating step. The tip is a closed off sharp cone. The exit hole is formed by grinding back the tip until a hole of the desired diameter is exposed.

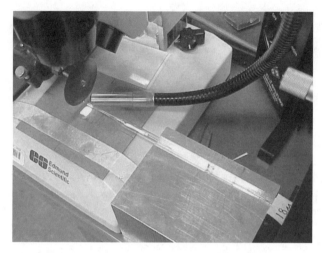

Figure 9.7   A grinding setup to controllably cut back the tip can be constructed using a V
groove alignment block to hold the pipette, a binocular microscope and a motorized
hand grinder spinning a 600 grit carborundum paper disc. The fiber optic coupled
illuminator directed at the end of the pipette is helpful but not absolutely necessary
for accurate grinding back of the tip.

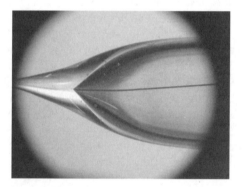

Figure 9.8   Microscope photograph of the closed conical tip of a pulled pipette. Fine gauge
tungsten wire is shown here pushed against the end to serve as an indicator of
grinding breakthrough.

the tip. Tungsten drift chamber wire is available in diameters down to 5 microns.
The way that the smaller diameter wires were used as grinding gauges was to attach
a short (~1–5 cm) length of it with tape or glue to a longer length of larger diameter
wire. Otherwise, it is impractical to handle. This wire gauge is pushed into the glass
tube and the drift chamber wire is pressed against the end of the sealed end. This
can easily be viewed though a binocular microscope (Figure 9.8).

   The wire breaking through the tip is the endpoint indicator for determining when
the grinding should stop (Figure 9.9). Another indication of breakthrough when
grinding back the tip under inspection by a microscope in real time is the introduction
of grinding dust into the interior of the pipette. Initial breakthrough as indicated by

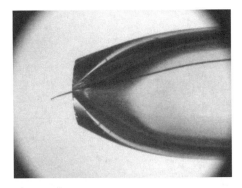

Figure 9.9   Tungsten wire emerging from the end of the pipette at completion of grinding back
the tip.

the presence of dust-like grinding debris inside of the pipette rather than wire
breakthrough is more often used by SLAC researchers as the stopping point prior
to microscope inspection of the tip when an extremely small hole is desired. Following the initial breakthrough, by alternating very short grinding intervals (i.e., 1
to 2 seconds) with a fine abrasive paper wheel on a motorized grinder and visual
inspection using a microscope with a calibrated reticle for inspection of the tip it is
possible in practice to get within ±2 microns of a desired hole diameter. Using these
methods it is practical to make ejection aperture holes with diameters as small as 5
microns in diameter.

No matter how one determines the grinding end point, the exact diameter of the
hole is obtained by directly observing the hole in the tip with a microscope with a
calibrated reticle (Figures 9.10 and 9.11).

The tip can then be flame-polished to reduce the sharpness of the edges of the
hole. Flame polishing the tip is a very tricky process. As the tip is heated to melt the
glass in order to polish the surfaces, the diameter of the hole also contracts and the
end starts rounding off. This can easily ruin the tip. The amount of heating necessary
to reduce the roughness of the aperture hole is very small. Passing the tip of the drop
ejector through the high temperature region of a propane hand torch for one second
is sufficient. Overheating the tip can close off the end of the glass nozzle tip.

The nozzle resulting from this type of forming operation has an extremely low
fluid impedance. One other advantage is that the dropper is highly chemically inert
compared with using modified inkjet printer hardware. The outstanding practical
advantage of this method is that it is a very inexpensive method for producing a
microdrop generator compared with those requiring micromachining. The disadvantages of this type of aperture are mechanical fragility and the virtual impossibility
of replicating the exact profile and hole diameter of the ejection aperture, making
each unit one of a kind. The mechanical fragility of the glass aperture makes it
difficult to clean due to the fact that physically wiping the aperture takes small chips
out of the edges of the ground glass hole. This asymmetrical aspect of mechanically
ground sharp-edged microholes makes the directionality of the ejected fluid a problem. An asymmetry in the hole on the order of microns can cause fluid jet deflections
on the order of tens of degrees. The tip is also delicate enough that simply touching

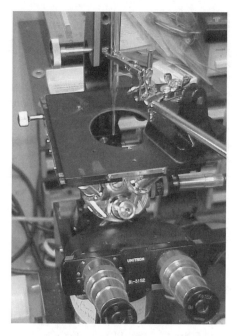

Figure 9.10 A microscope with a calibrated reticle is used to measure the diameter of the hole to determine when grinding back the tip should stop. The photo shows a Pasteur pipette with an thermally pulled ground back tip being inspected.

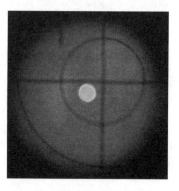

Figure 9.11 The 25-micron diameter hole in the tip of the glass pipette was photographed through the eyepiece of the microscope.

the tip against a hard object can destroy the aperture hole. Grinding the glass cone back to a precise prespecified diameter is a very delicate hand operation. One operational disadvantage to this type of constantly tapering profile is that any particle that gets into the fluid is guaranteed to eventually sink down to the aperture hole and clog it. This funneling property bug can be used as a feature, however, as it facilitates the use of a fine tungsten wire cleaning rod introduced from the reservoir end to unjam clogged apertures.

9.3 FABRICATING TAPERED PROFILE GLASS NOZZLES

The acutely angled edge nozzle pictured in Figure 9.9 has the advantage of low fluid impedance and is resistant to clogging due to its lack of a high aspect ratio channel where debris can wedge. Its operational disadvantages are that it is prone to misdirected jets due to the lack of a clearly defined preferential jetting direction and the sharp edges of the nozzle are prone to being easily damaged.

Switzer, in 1991,[2] outlined another method of making glass microejection aperture holes. This method utilizes the fact that heating a glass capillary tube causes its diameter to spontaneously contract. His process starts with small diameter (0.5 mm ID 0.7 mm OD) Pyrex capillary tube cut to a 19-mm length and squared off using 17-micron abrasives. This length of Pyrex is vertically rotated and heated with a small torch while being monitored with a microscope. The heating constricts the diameter of the inner bore while leaving the outer diameter unchanged The heating is terminated when the bore at the end of the glass tube has constricted to the desired diameter. At the end of this thermal constriction step, the tip has rounded thermally polished surfaces. The end is then ground flat using successively finer grades of abrasives. Switzer recommended going as fine as 1-micrometer abrasives. The nozzle that is produced by this process has a smooth taper to the final nozzle diameter with a sharp fracture free surface at the outer edge of the hole.

J.C. Yang et al., in 1997,[3] described another variation on thermally generated glass nozzles. Their paper described a process for making fluid ejection nozzles that utilizes larger bore tubing than Switzer. They mounted a length of 2.9 mm OD, 1.4 mm ID, tubing in a small lathe with both ends supported in rotating chucks. The tubing was heated while turning. When the glass softened, the tail piece chuck was pulled away, stretching and constricting the glass tubing without actually breaking it. When the glass cooled, it was cut at the desired locations, and the ends polished by a glass sander.

The processes described by Switzer and Yang,[2,3] have yielded nozzles less fragile and less likely to produce misdirected jets than those produced by the grinding back of sharp edge conical tips. The more nearly right angle cylindrical termination of the nozzle channel better defines a preferred fluid flow direction making these nozzles less likely to produce misdirected jets. Both Switzer and Yang[2,3] emphasized their observation that a smooth transition from the inner tubing bore diameter to the final nozzle opening diameter along with clean, fracture-free edges facilitated the making of satellite free drops. While smooth, symmetric ejection nozzle holes are less likely to produce secondary jets and satellites with fluids with poor rheologic characteristics, we have found empirically, though, that extremely nonsymmetrical, rough-edged ejection apertures can still produce satellite free drops when used in drop-on-demand mode with well designed fluids. Using fluids with the right rheological characteristics, we found that virtually any hole shape can produce satellite free drops. We found that fluid characteristics were far more important than the aperture geometry.

However, nonsymmetrical apertures, particularly those that have any kind of a height difference near the hole, can introduce very serious directional deflections of the fluid drops flight path. Research done at Xerox by Drews in 1991,[4] and Fagerquist

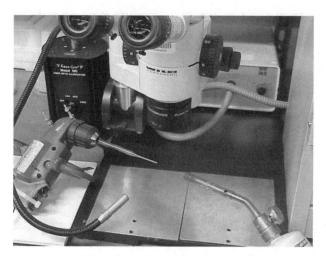

Figure 9.12 Fabrication setup for forming tapered profile glass nozzles. A propane torch is aimed towards the tip of a Pasteur pipette rotating at a rate one to two rotations per second. The torch flame should be set to a length where the tip of the hot blue flame tip barely extends out of the torch tip. For best results the heating operation should be as slow as is possible to allow for accurate determination of the point where the shrink down of the glass tip is optimal for the desired ejection hole diameter. The pipette needs to be rotated to avoid sagging and asymmetric shrink down during heating. An inexpensive way to mount and rotate the glass pipette during thermal forming is to utilize a variable speed battery operated drill. In our setup an external variable DC power supply is used to power the drill in place of the internal battery in order to more precisely set the rotation speed. A fiberoptic illuminator is used to optimize the illumination angle for observing the interior diameter of the glass pipette tip as it shrinks. An inspection microscope is used to determine the optimal heating rate and the heating termination point. The microscope should be selected for having adequate standoff distance from the torch to avoid damage to its optics. For efficient production in quantity a glass lathe with an electric heating element and a microscope with a calibrated reticle will reduce some of the trial and error associated with this simple forming rig.

in 1991[5] documented how small asymmetries in the fluid meniscus or aperture geometry height can cause very large, angular deviations in the directionality of the ejected fluid drops. As little as a few microns of vertical asymmetry near the ejection hole was capable of deflecting jets up to 30 degrees. Both asymmetric wetting and asymmetric aperture structure were able to cause this magnitude of fluid jet deflection.

Switzer's[2] method for making fluid ejection nozzles can be implemented relatively inexpensively and can be applied effectively not only to his small bore capillary tubes but also to the tips of Pasteur pipettes. Figure 9.12 shows a low cost setup for forming the ends of Pasteur pipettes into microdrop ejection nozzles.

The rounded thermally polished ends that are produced as a byproduct of the thermally initiated constriction of the pipette end is seriously suboptimal as an ejection nozzle tip because of the variable thickness frontal fluid layers that can accumulate in the depression at the tip. These variable thickness fluid layers can result in the need for large changes in the drive amplitude to produce appropriate jet speeds to produce monodisperse drops (Figure 9.13).

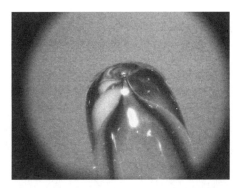

Figure 9.13 Glass pipette tip after initial thermal forming operation using the method of Switzer, 1991. The rounded thermally polished surfaces are not suitable as is for reliable operation when used as a drop-on-demand ejection nozzle. and need to be removed by grinding to form a nozzle face that allows a consistent fluid meniscus position. The glass tubing outside diameter is 1.5 mm.

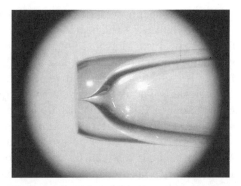

Figure 9.14 Finished nozzle using the method of Switzer, 1991. The glass tubing outside diameter is 1.5 mm. The nozzle tip hole diameter is 10 microns. This design for an ejection nozzle has a taper that results in a nearly cylindrical fluid channel at the nozzle tip. This is advantagous for ruggedness and obtaining consistent jetting directionality but offers a higher fluid impedance over a sharp edge nozzle such as that in Figure 9.9 and is more likely to clog.

The end of the tip is then ground back to remove the rounded region and to define the final diameter of the ejection nozzle hole. The grinding rig picture in Figure 9.7 is suitable for this task using a sanding wheel of 600 grit or finer carborundum paper. The gradual taper of the glass cone over that of the method outlined in section 9.2 makes it easier to grind to a predefined nozzle hole diameter (Figure 9.14).

9.4 SILICON MICROMACHINED EJECTION APERTURES

The hand-formed glass apertures have disadvantages because the precise hand work needed to form each aperture makes each ejection hole unique. This can cause problems for systems that require parallel, identically operating channels of inter-

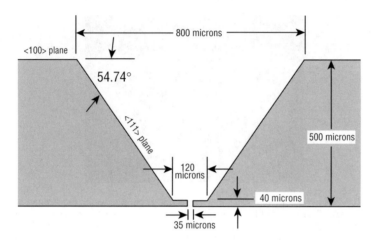

Figure 9.15 Cross-sectional profile of a silicon micromachined fluid ejection nozzle.

changeable spare units. The glass apertures fabricated by grinding a conical tip also are rarely perfectly axially symmetric, which produces offset jets that fire non-perpendicularly to the mechanical axis. Jets that fire at an angle have a secondary problem with accurately delivering microdrops, because the angle at which the fluid jet leaves the ejection hole is a function of the fluid velocity. A high-speed jet will be ejected nearly perpendicularly but as the jet velocity decreases, the angle of ejection will deviate in increasingly large amounts from a perpendicular jet. In extreme cases, we have observed fluid jets traveling nearly parallel to the plane of the ejection aperture surface. Glass aperture tips are also mechanically fragile, which limits the types of cleaning methods that can be used without damaging them. Physically wiping the tip will often fatally damage the hole. For these reasons, we have developed a process for micromachining ejection aperture holes using silicon micromachining (Figures 9.15 and 9.16).

The silicon microfabrication process we used to produce an approximation of this ideal shape had two principle process stages. The first process used chemical anisotropic to etch a deep pyramidal pit that came to a point, just short of the opposite end of the wafer. Directional ion etching was then used to form a cylindrical hole

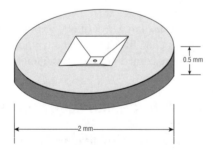

Figure 9.16 Fluid reservoir side view of the micromachined silicon fluid ejection nozzle detailed in Figure 9.15.

of a precise diameter that went from the tip of the pyramidal pit through the other end of the wafer. This process is similar to that detailed by Ernest Bassous of IBM in his 1974 patent[6] for using the anisotropic etching of silicon defined by silicon crystal planes for making fluid jet nozzles.[7–10]

The angle of the pyramidal pit is defined by silicon crystal planes to be 54.74 degrees with respect to the face of the wafer. A pyramidal pit forms when a silicon wafer cut with a <100> plane face is masked off with an etch resistant thin film such as silicon nitride with small openings and is placed in a caustic fluid such as potassium hydroxide. The <100> plane etches at a rate much faster than the <111> plane, so a four-sided, angular-walled depression starts forming. The etching process can be made self terminating by sizing the etching window such that the walls come together to a point at a desired depth in the wafer. If one wishes to create a structure that terminates in a plateau at a specific depth, the extent of the etching must be carefully monitored and manually terminated. There are a number of techniques to be discussed later to determine when to terminate chemical etching.

The optimal thickness of the silicon at the ejection hole side of the etched pit is determined by the diameter of the ejection hole. A nozzle hole length of one to two hole diameters is sufficient to suppress ingestion of air. Additional design factors are the plateau has sufficient strength to be resistant against being damaged by common cleaning processes such as being immersed in an ultrasonic bath.

For one design extreme, we fabricated an ejection aperture structure where the pyramidal pit etched down to the level of the 0.2-micron thick, silicon nitride masking layer at the other side of the wafer. This thin film was then penetrated with a plasma etcher. The 0.2-micron thick window actually survived the stress of fluid ejection but was destroyed after being placed in an ultrasonic cleaning bath.

It is possible to have the pyramidal pit come to a point and have the directionally etched cylindrical hole meet the pit at the tip. This is actually a mechanically stronger structure but for practical reasons, it is not always possible both to prespecify a cylindrical hole aspect ratio and etch the side walls down to where they meet at a point. The primary reasons are the variations in manufactured wafer thicknesses are much greater than the final lengths of the cylindrical ejection holes. I have encountered from the same manufacturer nominally specified 500-micron thick wafers that varied from 450 microns to 525 microns in thickness.

9.5 FABRICATION SEQUENCE FOR SILICON MICROMACHINED APERTURES

The process flow diagrams that follow are schematic representations of the actions of micromachining operations on a single structure in the silicon wafer. There can be from tens to hundreds of these structures being simultaneously manufactured.

Some of the processing details given in the descriptions are specific to the machines used in the Stanford Nanofabrication Facility. The details are necessary because they present some of the nonideal factors that can dominate the bulk of the time and money spent in processing silicon wafers to make micromachined structures. If one is unfamiliar with the basics of silicon microfabrication, a very readable introduction that includes a lot of the real world practical details of silicon micro-machining is: *Micromachined Transducers Sourcebook* by Gregory T.A. Kovacs, McGraw Hill, 1998. A short overview of the more common micromachining processes can be found in an article by Delapierre, 1989.[11] For historical interest, the pioneering article on silicon micromachining was written in 1982 by K.E. Petersen, Silicon as a mechanical material.[12]

The starting materials are 100-mm diameter, double-side polished, 500-micron thick, <100> faced silicon wafers. Though the digital devices chip makers use much larger wafers, 100 mm is currently the standard size used in micromachining facilities. Double side polished wafers are used because the process used to make these ejection apertures requires processing on both sides of the wafer. This is unlike making a microprocessor, for instance, in which only one side of the wafer needs to be worked with. These wafers cost about $25 each. It is best to buy all the wafers one will be using at one time from the same source before designing the photolithography masks because the thickness of the wafers is not well controlled. Wafers purchased that were nominally specified as 500 microns thick have varied from 450 to 525 microns in thickness. Within a given box, the wafers are usually within about 20 microns of each other in thickness. Since the wall angles of the pyramidal pits are defined by the 54.75 degree <111> crystal planes, the openings in the etch masking thin films must be sized according to the thickness of the wafer if the geometry of the ejection aperture at the ejection hole end is to be prespecified.

Two photolithography masks will be required. One mask will define the sizes and locations of the square openings made in the thin film, etch-resist layer. This will define the size of the pyramidal pits. The other mask will define the sizes and locations of the cylindrical ejection holes on the opposite side of the wafer. These masks can be designed and laid out with integrated circuit design software, which tends to be expensive and difficult on which to get up to speed, and frankly, is total overkill for the level of complexity of this task. The company that we had our photolithography masks made by accepted mask patterns laid out with AutoCAD. This is a lot more ubiquitous in most research lab settings. The masks physically are 127×127 mm quartz plates coated with chrome on one face. This chrome layer is selectively removed by computer controlled electron beam to create the pattern to be used to expose the photoresist layer on the wafer.

The number of ejection apertures to be placed on the wafer is determined by the mounting area needed to attach the wafer to the reservoir tube. It should also be kept in mind that wafers preferentially break and are thus diced along crystal planes, which constrain us to rectangular layouts for most economical fabrication processes. It is possible to use a deep directional ion etcher as a wafer saw to dice a wafer along noncrystal plane orientations, but this can be a very time consuming and hence expensive operation to specify as part of the fabrication process. This directional deep ion etching is a one wafer at a time operation and takes about 4 to 5 hours to etch through a 500-micron thick wafer. One of the factors that constrain the density of ejection apertures that it is practical to place on a single wafer is that the pyramidal pits both reduce the amount of silicon present and act as stress risers. Therefore, as the density of these pits goes up, the chances of the wafer breaking when placed on

Table 9.1 Typical Time Requirements and Simultaneous Wafer Processing Limits for Different Silicon Micromachining Operations

Operation	Number of Wafers	Time	Method
Box of wafers	25		
Chemical baths	25	15 min–2 h	Simultaneously processed in a 25 wafer cassette
Deposition furnaces	44	2–5 h	Simultaneously processed
Photoresist deposition	25	3 min/wafer	Individually auto processed from a 25 wafer cassette
Aligner (expose resist)	1	5 min/wafer	20 min setup, 1 wafer at a time hand loaded
Photoresist developer	25	3 min/wafer	Individually auto processed from a 25 wafer cassette
Plasma etcher	6	15 min batch	20 min setup, manually loaded
Deep chem etch	8	12 h	Simultaneously processed in small bath
Directional ion etch	1	1–2 h/wafer	One wafer at a time, hand loaded

a vacuum chuck or transfer belt increases. As a compromise, we used a 1-cm grid spacing for our ejection apertures. This also helped reduce layout errors because the coordinates of the structures could be made to lay along easily verified locations with easily incremented numerical coordinates.

The processing of these wafers to make ejection apertures involves a mixture of operations that are both serial and parallel in nature. For instance, when cleaning the wafers in a chemical bath or growing thin films in a deposition furnace, it makes essentially no difference in the machine time whether there is one or a hundred wafers in the acid bath. On the other hand, there are a number of machines, such as the deep reactive ion etcher, that only accept one wafer at a time and require hand loading of each wafer. Other operations are such that there is a certain fixed amount of setup time needed, but the amount of human intervention needed is the same for a hundred wafers as one, although the machine time consumed is linear with the number of wafers. This is the case for the photoresist coaters and developers (Table 9.1).

Unlike processing silicon wafers for electronic device manufacture in a silicon foundry, the micromachining of structures in research facilities includes steps in which the advantage of parallel processing of large quantities of wafers simultaneously is lost. In particular, the directional ion etching process that generates the ejection hole utilizes a machine that can only accommodate one wafer at a time. There are also processes that require long periods of time and can only accommodate a limited number of wafers. For instance, the chemical etch used to create the pyramidal pits is a very time consuming 12-hour process in which only eight wafers at a time can be fit into the chemical bath at the facility (Stanford Nanofabrication Facility) where the wafers were processed.

One other factor to be considered is whether one would even want to automatically batch process a large amount of wafers at one time. Microfabrication facilities are often associated with major teaching universities and are used by their students for instructional purposes, as well as by seasoned in-house technicians and industry engineers for production prototyping. In this kind of user environment there is a very real possibility that the processing machine could be put seriously out of calibration, be damaged or contaminated by an inexperienced previous operator.

Wafer processing in a combined instructional and research environment usually involves working in a facility in which the machines cannot be depended upon to behave identically to what one experienced a few months or even a few days in the past. Even if the process one is using has been debugged, the misbehavior of a given machine can trash a processed batch of wafers if one simply loads the wafer cassette up and sends them through.

A micromachining lab typically produces a lot of limited quantity processing runs using unique machine settings in a chaotically scheduled environment with many machine users of highly variable amounts of experience. This is fundamentally different from a commercial silicon device foundry where a particular operation to be done to a wafer is performed by a dedicated technician on a given machine and is the same as what has been done previously to nearly identical wafers for the last several months or years, except possibly for the mask pattern. There is a certain amount of reliability that one can count on that comes from constant operation under similar conditions of machines run exclusively by dedicated experienced technicians. In contrast, in a research-oriented micromachining environment one usually runs no more wafers simultaneously than what one can afford to lose in terms of dollars and time if the batch turns out to be fatally damaged by a misbehaving processing machine. One should use caution when processing high value wafers in shared micromachining facilities through machines that one has not personally run test wafers through immediately prior to actual use.

9.5.1 Processing Steps for Fabricating Silicon Fluid Ejection Nozzles

Step 1 — The first step is to deposit on the wafers a thin film layer of a chemical etch mask material resistant to the chemicals able to form pyramidal pits by deep, silicon crystal plane specific etching (Figure 9.17). This process is performed in a high temperature oven in which various gases are introduced, which then react with the surface of the silicon wafer to grow a thin film layer of silicon nitride (effective mask for both potassium hydroxide (KOH) or tetramethyl ammonium hydroxide (TMAH) etching) or silicon dioxide (effective for masking a TMAH etch). Not shown on the process sheet is the cleaning stage that proceeds this furnace operation.

The wafers must be cleaned sequentially in a heated sulfuric acid-hydrogen peroxide bath followed by a hydrochloric acid bath finishing off with a hydrofluoric acid bath to remove the thin oxide film that forms on exposure to air. The wafers are then rinsed in deionized water and spin dried. This operation must be done just prior to the furnace operation. The wet etch chemicals typically used to form the pyramidal pits are potassium hydroxide or tetramethyl ammonium hydroxide (TMAH). A silicon nitride (KOH or TMAH) or silicon dioxide (for TMAH) layer is needed because photoresist is unable to withstand immersion in these silicon wet etchants. Silicon nitride was chosen as the masking material because of its superior contact angle and mechanical durability over silicon dioxide when acting as the surface of an ejection aperture. The wet etch proceeds at a rate of about 1 micron

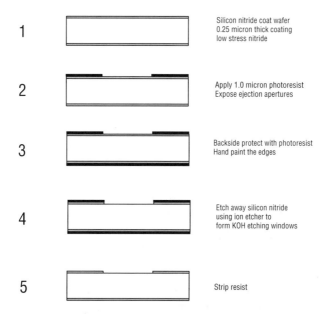

1 — Silicon nitride coat wafer
0.25 micron thick coating
low stress nitride

2 — Apply 1.0 micron photoresist
Expose ejection apertures

3 — Backside protect with photoresist
Hand paint the edges

4 — Etch away silicon nitride
using ion etcher to
form KOH etching windows

5 — Strip resist

Figure 9.17 Process flow diagram for the deposition of a silicon nitride thin film chemical etching masking layer on to a silicon wafer. The processing steps for one of up to hundreds of identical structures is represented in schematic fashion. The diagram is not to scale. Some of the thin film coatings are less than a micron in thickness.

per minute. In principle silicon nitride is virtually insoluble in KOH so a thin film layer that is only a few tens of manometers thick can screen off the parts of the wafer that one wants to protect.

In practice, it is prudent to make a much thicker layer so that mechanical abrasion from wafer handling does not damage the film and cause etching in undesired areas. The thickness of the layer determines its color via interference film effects so that a sight variation in film thickness can be used to color code different manufacturing runs. The thickness of the nitride layers used by SLAC researchers ranged from 0.2 to 0.25 microns.

It is important that the nitride recipe used forms low stress thin film. High stress nitride is more prone to opening up surface defects during etching and producing pinholes, isolated etched through regions on the face of the wafer, and etching notches at the edge of the wafers that act as crack promoting stress risers. The nitride forms at a rate of about 40 to 50 Angstroms per minute. There is also a manual loading and unloading operation plus the pump down and gas purging cycle at the start and end of the deposition cycle that adds about two hours to the operation in addition to the required thin film formation time.

Step 2 — The next step is to apply a layer of photoresist to the wafer so that the opening for the chemical wet etch can be cut away by a plasma etch through windows in the photoresist. A standard process can be used at this stage with the wafers loaded into the cassette. It is best to not coat the wafers with photoresist until just prior to the time when they will be exposed and developed. This is because it is possible for the storage box that the wafers are kept in to be exposed to normal room lighting and have the photoresist compromised.

Prior to having the resist applied by the photoresist spin on machine, the wafers have to be singed to remove surface-absorbed atmospheric moisture. This simply involves placing them in a designated 150 degree oven for 30 minutes and then letting them cool down to room temperature prior to coating. There is a one-hour window between removing the wafers from the singe oven and coating them.

The photoresist is then exposed using a contact aligner. This is a machine that takes the photolithography mask, allows alignment with existing structures on the wafer, and then places the mask in contact with the wafer and illuminates the mask with ultraviolet light allowing the features defined by openings in the mask to be exposed to UV light. The photoresist most commonly used is positive resist. When using positive resist, areas exposed to light are removed by the development process.

Since there are no existing structures at this point, the only alignment criterion is that the orientation of the etching windows be parallel to the crystal axis of the wafers, the accomplishment of which depends upon the particular machine one is using. The flat indents formed into the sides of the wafers define the crystal planes. A diamond scribe can be used to form a line across a special test wafer parallel to the flats. The wafer holder usually has an alignment feature that can align to the wafer flats. With the test wafer with alignment scratches in place in the holding jig, the relative rotational orientation of the wafer and mask are adjusted until the crystal planes align with the orientation of the windows in the mask. A misaligned rotational orientation with the wafer crystal plane will result in a larger than designed for pyramidal pit. After exposure, the wafer can be developed using the standard recipe on the automatic machine. It is usually prudent after exposing one wafer to develop it and inspect the resulting pattern prior to exposing the rest of the wafers.

Step 3 — The photoresist spinner deposits photoresist only on the upper surface of the wafers as it is held by a vacuum chuck. This leaves the sides and the back of the wafer unprotected against the effects of the plasma etch.

This lack of complete photoresist coverage is normally irrelevant for device processing since a few microns lost off the edge or back of the wafer is unimportant. This is not the case for an operation meant to form a protective layer against a wet etch operation capable of etching through a wafer. To backside protect a wafer, a layer of photoresist is spun onto it. This process is similar to the operation used to deposit the front side layer. The procedure is modified though by the requirement to protect the photoresist already applied to the front of the wafer and the lesser quality requirements for photoresist that are not required to be exposed and developed.

For the backside protect process the following procedure is used:

- The singe step is not needed.
- The vacuum chuck is wiped down with acetone to remove residual photoresist.
- The photoresist deposition machine is used in manual mode.
- The priming step and heat plate bake stage is bypassed.

- The process is triggered (i.e., hit start button) with an empty cassette in the machine.
- A wafer is placed by hand using tweezers onto the feedbelt going to the vacuum chuck spinner.
- After the photoresist is spun on and the chuck releases the wafer, the wafer is taken with tweezers and placed in a Teflon cassette.
- After all of the wafers are processed the Teflon cassette is placed in the 110-degree post bake oven for 30 minutes to harden the resist and evaporate away the resist solvents.

The backside protection unfortunately does not insure that there will be photoresist covering the edges of the wafers. Gaps in the photoresist on the edge of the wafer will produce nicks in the edges of the wafer, acting as stress risers and making the wafers highly vulnerable to breaking at those points. I have seen wafers with those edge defects break upon being placed in a vacuum chuck or upon hitting a transfer belt end stop plate. The edge can be protected by painting photoresist on the edges with a small paint brush. To perform this operation, the solvent wet bench is used. The wafers are supported while they are being painted by being placed on Pyrex beakers. After being painted, the wafers are placed in the 110-degree oven, again being supported by Pyrex beakers.

Step 4 — A plasma etching chamber using SF_6 and CF_3Br gases removes the silicon nitride thin film from the regions exposed by the photolithography operation. The principle cautions about this process are not to assume that the wafer will etch evenly over its entire surface or that each of the wafer holders will have the wafers positioned in them etch at identical rates. It is advisable to deliberately underetch the first batch to check on these factors and to verify the machine settings before attempting 100% thickness removal of the nitride. The silicon nitride is removed at a rate of about 800 Angstroms per minute while the resist is removed at a rate of 540 Angstroms per minute. Since the standard photoresist coating is one-micron thick this gives a healthy operating margin.

Step 5 — Since the resist is on all surfaces of the wafer and not just on the face, it is most conveniently stripped in a piranha etch (H_2SO_4 H_2O_2) bath. The principle caution here is to lower the wafer-filled cassette slowly into the bath. This is because of the possibility that gas bubbles generated by the dissolving photoresist when it first hits the piranha etch will float the wafers out of the cassette. This resist strip process takes about 20 minutes after which the wafer filled cassette is dump rinsed in deionized water and spun dried.

Step 6 — This step is one of the most time consuming due to the length of the time required to etch through the wafer plus the amount of stopping and testing and restarting required to hit the precise etching depth end point (Figure 9.18).

The general setup for this operation is to place a quartz beaker containing the caustic etching compound, usually KOH or TMAH, in a temperature controlled water bath with a water cooled condenser covering the top of the etching beaker. The etching rate is a strong function of temperature and a lesser function of the concen-

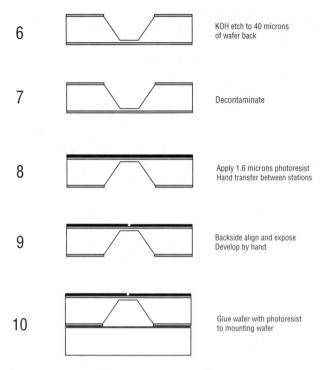

Figure 9.18 Processing step sequence for chemical etching of the pyramidal pit and preparation steps for ion etching of the cylindrical orifice hole.

tration of the etchant. This makes manual monitoring of the rate of etching essential for reaching a desired end point. The most reliable way of monitoring the depth of etching is to periodically remove the wafers from the etching bath, rinse them in deionized water and then measure the depth of the pits using a microscope with a depth calibrated focusing stage. At high magnification with one of these microscopes one can readily resolve the depth of etching to plus or minus a few microns. A wafer thickness gauge or a set of precision calipers can be used to measure the wafer thickness to find the thickness of the lower plateau region of the pit.

If one wishes to stop the etching at a precise point in order to have a predefined ejection hole channel length, the final stage of the etching can be very time consuming. The wafers must repeatedly be pulled from the etchant, cleaned and dried, and then optically inspected. It is important to check the pit depth at different locations on the wafers as well as on different wafers to be sure that temperature gradients have not been set up in the etching bath that are causing different rates of etching at different points in the bath.

It is important to verify that the micromachining facility has strict rules regarding material cleanliness requirements for wafers to be processed by subsequent machines. If wafers are found to have been etched in a nonquartz beaker or in a quartz beaker that had been used with wafers containing gold, the wafers that were etched may not qualify for being processed by the machines that must be used to finish off the process. Be sure that the beakers, wafer cassettes and tweezers one uses are made of approved materials and have not been contaminated by a prior user.

Step 7 — The wet etchants used to form the pyramidal pits use compounds that contain KOH or are assumed to be contaminated with TMAH rare Earth (potassium, sodium) ions. These elements can poison MOS transistors and other active devices. In order to qualify for processing in most ion etching machines, these contaminants must be removed. Fortunately there are effective procedures for decontaminating the wafers from these particular elements, which involve properly sequenced multiple rinses in deionized water and acid baths. This decontamination procedure can be quite time consuming (several hours) if it involves the use of the bench top acid baths due to the time required for heating and cooling and exchanging the solutions. The approved procedure differs in each lab and, in my experience, within each lab at different points in time (Figure 9.18).

Step 8 — The function of the next set of operations is to form the cylindrical ejection aperture hole in the opposite side of the wafer that was the processed to form the pyramidal pit. The wafers should be inspected for edge defects after etching to determine if they are intact enough to be placed in automatic processing machines. Nicks in the edges of the wafers can cause them to crack if they are driven into mechanical end stops by drive belt transfer stages. If the edges of the wafers are clean, they can have photoresist deposited on them in a manner identical with that of a conventional wafer. If there are edge defects, then the transfer stages should be bypassed and the wafers moved from station to station on the photoresist deposition and developing machine using tweezers.

Step 9 — The photoresist must now be exposed using the mask that defines the ejection aperture hole diameters. This must be done on an aligner that permits what is commonly referred to as backside alignment, that is alignment of a pattern on one side of the wafer with respect to existing structures processed into the other side. A thicker coating of resist is needed at this stage than for the initial plasma etch. In this operation, the ion etcher must penetrate on the order of 50 microns of silicon. The silicon etches at a rate of about 1 micron per minute. The photoresist etches at a rate of about 0.4 microns per minute. This gives an etching rate ratio for the photoresist to silicon of about 25 to 1. To effectively mask off the wafer for this operation, the thickness of the photoresist must be a minimum of 1/25 times that of the depth one wishes to etch. If the wafer has gone through the chemical etch without damaging the edge, then developing the exposed photoresist can be done with the automatic developer. If there is edge damage, the wafer can be developed in a shallow beaker, timed by hand with a stop watch.

Step 10 — For processes where the wafer is pierced all the way through by the deep directional ion etcher, the wafer containing the micromachined structures must be mounted on another wafer to protect the wafer chuck. The most common way to adhere one wafer to another, in a way that does not introduce material contamination into the ion etching chamber, is to use photoresist as a temporary glue. There is a procedure in the literature that utilizes weights and a hot plate to assist in adhering two wafers to photoresist but, in practice, this method ended up being unnecessarily complex. Fresh photoresist is tacky enough that simply resting one wafer on top of another that just had photoresist spun onto it is sufficient to glue them together.

The following procedure is the one SLAC researchers successfully used:

- Using manual mode, coat backing wafer with a 7-micron thick layer of photoresist. Remove wafer from spinner as soon as the resist has been spun on. Singing, prepriming, and hot plate baking are not needed.
- Pick up the wafer from the vacuum chuck using tweezers and place the wafer on top of a glass beaker of a diameter less than the wafer diameter photoresist side up. Immediately place the to-be-processed wafer on top of this mounting wafer and align the edges of the wafers flush with each other. There is a 20-second window to accomplish this after the wafers have come into contact.
- Take a Teflon wafer cassette and rest it on end so that the wafers placed in it sit horizontally. Use this to hold the wafers after the photoresist has taken its initial set.
- After all wafers have been processed, place the Teflon cassette containing the wafers into a 110°C oven for one hour.

Step 11 — Directional reactive ion etching is used at this stage to produce the cylindrical ejection hole. A directional ion etcher is a special dry plasma etching machine that is capable of very high aspect ratio cutting away of silicon in the manner of an electric drill or milling machine. This is in contrast to conventional plasma etching chambers where the cutting away of the material is nondirectional, producing hemispherical cavities radiating from the aperture opening. Overetching is not a problem for this process, unless the photoresist thins out enough that the face of the silicon aperture starts etching away (Figure 9.19).

Step 12 — The easiest most reliable method of separating photoresist glued wafers is to utilize the photoresist stripping bath (hot H_2SO_4 and H_2O_2).

- Immerse the wafer in the stripping bath for 2 hours. Insure that the fluids are fresh, spike the bath with additional peroxide if necessary.
- Remove the wafer cassette and dump rinse them in deionized water. Do not spin dry them! The semidissolved photoresist debris that is still between the wafers will contaminate the spin dryer. Blow dry the wafers in the cassette over absorbent wipes.
- Gently slide the wafers apart from each other. If they do not easily slide apart, put the wafers back in the resist strip bath. Do not attempt to pry them apart or they will crack.
- Place the separated wafers back in the wafer cassette and immerse them again for a standard strip time interval (~ 20 min) into the resist stripping bath to clean off the liquefied resist residue.
- Dump rinse and spin dry.

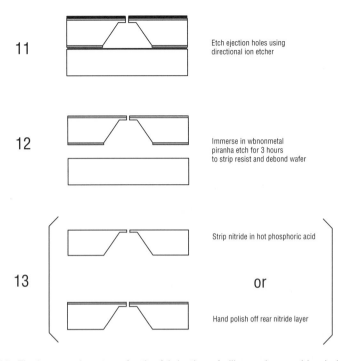

Figure 9.19 Final processing steps for the fabrication of silicon micromachined ejection nozzles.

Step 13 — The purpose of the last step is to remove the silicon nitride from the fluid reservoir side of the wafer. This is necessary because glass welds very poorly to silicon nitride. The easiest way to strip off the silicon nitride while leaving the silicon structure intact is by immersing the wafer into a hot phosphoric acid bath. The disadvantage is that this would remove the silicon nitride from the front surface also. Silicon nitride is a better front aperture material than bare silicon because it has a larger contact angle with the fluids most commonly used for microdrop applications and it is very hard and abrasion resistant. The method we currently use is simply to sand the silicon nitride layer off the back of the wafer with 600 grit carborundum sand paper.

The wafer is now ready to be diced with a wafer saw.

9.6 ELECTROFORMED EJECTION APERTURES

Metal ejection aperture plates can be microfabricated using electroforming equipment far less expensive than that used for silicon micromachining. One example of a fabrication sequence for a stainless steel and nickel ejection aperture plate is in Figure 9.20.[13,14] The process shown relies upon the electroplated nickel being usable as a chemical etch mask for the steel substrate. In principle, any pair of metals in which the electroplated metal can act as an etching mask for the substrate plate

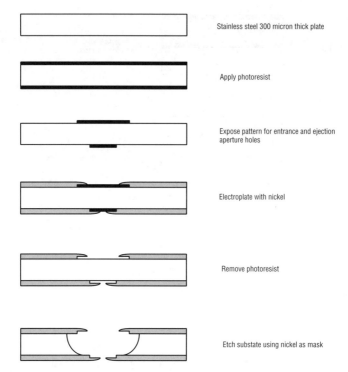

Stainless steel 300 micron thick plate

Apply photoresist

Expose pattern for entrance and ejection
aperture holes

Electroplate with nickel

Remove photoresist

Etch substate using nickel as mask

Figure 9.20 Processing sequence for the making of metal electroformed fluid ejection
nozzles.[13,14]

can work. The main virtue of this process over the previously detailed silicon based
process is simplicity and low cost. The main disadvantages are that a bimetallic
ejection aperture is not as chemically resistant to corrosion and abrasion as a silicon
structure, and that the final ejection hole diameter is a function of the plating time
and condition of the plating bath, as opposed to being determined by a photolithography mask pattern.

9.6.1   Jewel Bearings and Nozzles

A source of rugged fluid ejection nozzles suggested by Steven Zoltan in his 1972
patent disclosure[15] was precision-machined jewel bearings, typically made from
ruby, sapphire, diamond, and tungsten carbide. Though originally intended as bearings, these precision-machined perforated discs can be used as drop-on-demand
ejector nozzles. In addition, some of the companies that manufacture these precision
jewel bearings also market a line of ruby and sapphire components specifically
intended for use as continuous flow fluid jet nozzles. A jewel nozzle made from
sapphire and ruby surprisingly can be thermally welded to Pyrex tubing in a manner
identical to that used to attach micromachined silicon nozzles to glass tubing.

The diameters available for the ejection hole are about 10 to 100 microns and
larger. Presently, there is a serious aspect ratio nonoptimality for most of the com-

mercially available jewel nozzles and bearings. The holes are far longer with respect to their diameters than is necessary for reliable fluid drop ejection. As a result, the drive amplitudes are much larger than for a properly designed micromachined ejection nozzle. There are some companies that make jewel bearings with conical starter holes that are more optimal but the diameters of the through hole section is restricted to holes the order of 50 microns in diameter and larger. This minimum machinable diameter will probably shrink with time.

Some manufacturers of these products are: Swiss Jewel Company and Comadur SA in Switzerland; Gardella in Italy; Tecpro in Australia; Moser Company and Imetra in the U.S. The price per jewel nozzle is in the $20 to $100 range.

Swiss Jewel Company
437 Chestnut Street
613 Lafayette Building
Philadelphia, PA 19106
www.swissjewel.com
215–925–2867

Comadur S.A.
57, rue Girardet
Le Locle, CH-2400 Switzerland
www.comadur.ch
+41 32 9308311

Gardella s.r.l.
via Fleming n° 1
I - 27058 Voghera (PV)
www.gardella-srl.it
0039/0383.41575

Technical Projects (Tecpro)
4/44 Carrington Road
Castle Hill NSW 2154
Australia
www.tecpro.com.au
(02) 9634 3370

Moser Company
518 State Route 57
Phillipsburg, NJ 08865–9484
www.mosercompany.com
908–454–1155

Imetra
Cross Westchester Executive Park
200 Clearbrook Road
Elmsford, NY 10523–1396
www.imetra.com
914–592–2800

9.7 METAL OPTICAL PINHOLE APERTURES

One readily available inexpensive source of precision small diameter holes is the stainless steel foil optical pinhole. However, after being tested as ejection apertures for tubular fluid reservoir drop ejectors, researchers found that they performed very poorly.

The ones we obtained were 0.5 mil (13 microns) in thickness. The electrical drive amplitude for fluid drop ejectors made with these apertures was about ten times that of drop ejectors made with silicon micromachined apertures. We speculated that this may be due to the flexibility of the thin foil apertures absorbing the pressure impulse. These apertures also ingested air into the fluid reservoir each time they went through a drop ejection cycle. This is due to the fact that the foil type ejection apertures lack a refill column combined with the high pressure pulse amplitude needed with this type of aperture to produce drop ejection.

Once drop ejection took place it was also unstable. This was using a 75% propylene glycol, 25% water mixture that ejects without any problems from drop ejectors with glass or micromachined tips. We have had thicker 0.2 mm stainless steel apertures fabricated using laser cutting of a conical ejection holes. These worked better than the foil apertures, but required a higher amplitude for ejection due to the smaller cone angle for the particular apertures that were furnished.

9.8 SMALL HOLE MANUFACTURING TECHNOLOGIES — A SUMMARY OF TECHNIQUES

The hard part about making fluid jet ejection nozzles is making precise small diameter holes. There are other methods for making holes for fluid ejection apertures with varying degrees of suitability for experimental laboratory use. Some may be unsuitable because of the cost of development of the necessary details of the process, some because of restrictions on the materials that can be perforated and their limited chemical compatibility with the fluids one may wish to eject. Other techniques for making small holes may be unable to make conical, tapered holes needed for optimal operation with minimum pulse energy.

Don Hayes and David Wallace of MicroFab Technologies published the definitive paper on techniques for making small holes titled appropriately enough: *Overview of Small Holes*.[16] This paper can be a source of creative ideas for alternative ways of fabricating fluid jet ejection apertures. The following is a summary of some of the techniques they listed for making small holes.

9.8.1 Mechanical Drilling

Drill bits down to 15 microns in diameter have been manufactured. Production mass drilling of PC boards with mechanical drill bits with diameters down to 100 microns is routine in industry. At the time of the paper's publication (1989), mechanically drilling 10 micron class holes through hardened materials, such as silicon wafers, was beyond what had been reported as being technically possible. Since then, there has been a significant advance in the ability of to drill small holes through hardened materials with dimensions relevant to the making of microdrop fluid ejection nozzles. The existence of jewel bearings with 10-micron diameter cylindrical holes implies that the technology has been developed to make this microdrilling task available on a routine production basis.

9.8.2 Mechanical Punching

Mechanically generating a hole by pressure shearing using a hardened punch has been done for hole diameters down to 25 microns in diameter. Hole aspect ratios of 0.5 to 5 have been achieved in practice. Diameter and aspect ratio limitations currently eliminate this method from being applicable to the generation of sub-10-micron diameter holes in inkjet orifice plates.

9.8.3 Photosensitive Glass

This method of small hole generation utilizes a special Dow Corning glass (Photform®, Fotoceram®) that selectively crystallizes where previously exposed to UV light when heated in a furnace. These crystallized regions then can be etched with special solvents at a 15 to 20 times greater rate than the uncrystallized glass. Hole sizes down to 25 microns have been achieved using this technique.

9.8.4 Soluble Core Glass Fibers

This hole generating technique has been used to routinely generate very high aspect ratio near perfectly cylindrical profile holes down to 5 microns in diameter by embedding sacrificial etchable material into the material wafer at the time of manufacture. The technology behind this was developed primarily to manufacture microchannel plates for electron amplification for image intensification vacuum tubes. This method is inapplicable for making microdrop ejection nozzles for economic reasons, as well as being unsuitable for making noncylindrical hole profiles.

9.8.5 Laser Drilling

Lasers are capable of drilling holes in virtually all solid materials. The drilling process is in general serial, but has been adapted with reflective metal masks over plastic films to be used as a parallel micromachining method. Parallelizing the laser drilling process obviously is dependent upon there being a masking material relatively much more resistant to optical ablation than the material to be penetrated. Masking is unlikely to be usable for drilling into silicon, though stainless steel masks have been used to make arrays of micron-sized holes in polyamide. Laser drilled holes through thick materials are usually conical in taper, which is desirable for ink jet orifices. This taper can be controlled by the laser drilling process parameters such as the energy per shot, the number of shots per hole, the degree of focusing and the pulse duration. Except that this hole generating process is serial, and that precision laser drilling apparatuses are not universally available in IC processing facilities, this can be a very applicable technology to cut the conical profile holes needed for the droplet generators.

9.8.6 Electroforming Over a Sacrificial Post

This is a widely used method in industry for forming ink jet orifice plates. In this process, photoresist is used as a masking and sacrificial material to define front and back hole diameters by limiting the electrodeposition of a metal film on both sides of the orifice plate. Nickel is the most often used metal. The photoresist is then removed and the nickel films at either side of the wafer are used as etching masks for a chemical etch to penetrate the substrate, connecting the front and back nickel orifices. This method has been used to electroform orifice plates with holes down to 10 microns in diameter. The use of this process depends on very tight control over the electroplating steps, since the final diameter of the holes is determined by the precise timing of the overgrowth of the plating over the photoresist mask. Much trial and error is probably required to develop a particular process using this technology.

9.8.7 Photon Directed Chemical Etching

This is a form of enhanced chemical etching in which an electrical bias and optical illumination is used to speed the etching in a highly anisotropic manner such that very deep, high aspect ratio holes can be formed. This is distinguished from laser enhanced etching in that a nonthermal enhancement mechanism is used based on the selective production and directing of holes (positive Si lattice charge carriers in this case) in the wafer. This method has the disadvantage of being a nonstandard fabrication technique not readily available in commercial production facilities.

9.8.8 Electron Beam Machining

This technique uses material ablation by electron beam impact to generate a hole in the target material. The material removal mechanism is thermal in nature, similar to laser material ablation.

9.8.9 Electro-Discharge-Machining (EDM)

EDM utilizes a spark discharge between a cutting electrode and the material to be drilled through to remove material. The electrical excitation is typically supplied as a series of pulsed discharges. The cutting takes place with the workpiece submerged in fluid. Holes as small as 15 microns have been cut through hardened metal with this technique. Micro EDM cutting, in principle, can be used to cut conical profile holes with exit orifices in the 10-micron diameter range.

9.8.10 Lithography and Chemical Etching

This is the standard IC fab technology of utilizing photoresist to act as a masking layer for an etching agent intended to chemically dissolve the material to be removed. Alternately, the photoresist can be used to define a region where masking material is laid down. Many different etching agents, processing schemes and methods of achieving various degrees of anisotropy are possible. Hole sizes down to the sub-micron-diameter range is possible, in principle, though high aspect ratio small diameter holes are difficult to manufacture in practice.

9.8.11 Lithography and Anisotropic Reactive Ion Etching

This is distinguished from chemical etching from the use of an externally applied electric field enhancing the removal of material over what would be achievable with activated chemical reactants alone. Very high aspect ratio etching independent of crystal planes is possible. One practical difficulty for making deep etched holes using many of the high aspect ratio etching processes is that the masking material is removed at very nearly the same rate as the wafer material. However, very precisely defined micron-diameter range holes are possible with this technology.

9.8.12 Thermally Tapered Glass Pipettes

This is a widely used way of making small diameter holes for ejecting fluid microdrops. A glass tube is locally heated and then pulled such that the heated section tapers to a closed tip and breaks off at that point. The point is then ground back until a hole of the desired diameter is produced. The great advantage of this approach is that no costly or sophisticated hardware is needed to make this type of ejection aperture. The disadvantages are the fragility of the ejection aperture tip, the skilled labor intensive nature of the process and the difficulty in obtaining an exact reproducible desired hole diameter and profile for large production runs.

9.9 ANTI-WETTING SURFACE COATINGS

Wetting of the orifice surface can seriously compromise the operation of drop generators for multiple reasons. The first is that excess fluid buildup can suppress ejection outright. Another issue detailed in Drews[4] and Fagerquist[5] is that asymmetric fluid build up on the surface of the ejection orifice can deflect the direction of drop ejection by tens of degrees. Negative internal pressure is one technique used to reduce ejection orifice fluid build up around the ejection hole. Another technique in common use is the deposition of anti-wetting thin films.

Anti-wetting surface films reduce the fluid accumulation over the aperture by modification of the fluid-surface contact angle. Contact angle is a way to quantify the tendency for a fluid to wet a given surface as opposed to cohering to itself. A high contact angle signifies a surface that repels the fluid so that the fluid drops on this kind of surface form small mobile spheres.

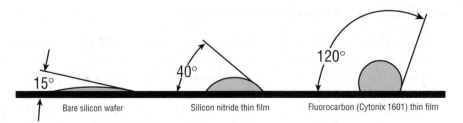

Figure 9.21 Contact angle changes for water drops in contact with different surfaces.

Fluoropolymers (e.g., Teflon) are the most common high contact angle coatings for aqueous fluids. A low contact angle would be measured for a fluid that easily wets the surface. Fluid drops on this kind of a surface would wet the surface forming a thin film spread out over the surface, as opposed to forming self-contained balls. Commercial windshield coatings such as RainX® (Pennzoil-Quaker State Company, P.O. Box 2967, Houston, TX) alter the contact angle of the glass so that water forms thin sheets rather that small drops. A fluid drop placed on the boundary between a high and a low contact angle surface would jump over to the low contact angle side leaving the high contact angle surface dry (Figure 9.21).

This property can be used to suppress wetting of the ejection aperture surface by coating the surface with a high contact angle thin film while having the interior of the drop ejector coated with low contact angle materials. Any residual fluid that ends up over the ejection aperture hole would have a natural tendency to be returned to the interior of the drop generator. This tendency assists the negative internal pressure applied by the manometer since all of the fluid has more of a chance to return without part of the fluid breaking off from the main mass and adhering to the ejection orifice surface. For some fluid, the application of a high contact angle thin film can eliminate the need for an external manometer to supply negative internal pressure.

The measurements in Tables 9.2 to 9.4 were roughly gauged by eye and should be taken only as indicative of relative affinity of different surfaces to distilled water. Tests of this type can be used to rank order and roughly gauge the suitability of different fluids and microdrop generator contact surfaces. For instance, a fluid-surface combination such as silicon oil on glass resulted in the oil not forming a discrete drop at all, but instead formed a constantly spreading thin film. In contrast on Teflon surfaces, the silicone oil did form cohering distinct drops as opposed to a constantly spreading thin film.

This corresponded with our observations when using silicone oil in fluid drop ejectors with glass or silicon ejection apertures. No matter how much negative internal pressure was applied, a thin film of silicone oil a few microns thick (made evident from Newton's rings) would spread from the ejection aperture hole. Our experience, in fact, was that silicon oil was very difficult to obtain reliable drop-on-demand from and had very difficult restart characteristics due to the variable fluid meniscus that tends to accumulate over the ejection hole. This is in contrast with water, where either Teflon coating of the aperture or a slight amount of negative pressure was sufficient to prevent any detectable fluid leakage.

9.9.1 Coating Methods

We have used three methods of obtaining thin film Teflon coatings over our ejection apertures.

The first method was to have the thin film applied as the part of the microfabrication process. Thin films applied in a plasma deposition chamber are optically smooth and highly uniform in thickness. The major disadvantage to this technique is that thermal welding of the aperture to the fluid reservoir tube is not possible due to the limited temperature tolerance of the Teflon thin films. One practical issue associated with this method is that the machines able to perform this process are typically very costly, highly maintenance-intensive and are only found in integrated circuit and micromachining laboratories. Many of these machines have strict rules concerning the prior history of items allowed to be processed due to contamination issues. For some machines, the operating rules dictate that if the item was ever touched with ungloved hands, the item would be barred from ever being introduced into the processing chamber. Teflon coating assembled microdrop generator assemblies inside of these plasma thin film deposition machines is usually out of the question. Our ejection apertures had to be fabricated totally within the IC lab including the thin film coating. Recoating a drop ejector that had its Teflon thin film damaged by improper wiping, for instance, is not possible.

As an alternative, there are companies that manufacture fluids that can be used to apply Teflon thin films to objects using a dip, spray, or brush-on applicator. DuPont (DuPont Building, 1007 Market Street, Wilmington, DE) manufactures a line of Teflon coating fluids based on micron sized solid Teflon particles suspended in a volatile carrier fluid. This line of Teflon fluids go by the commercial trade name of Krytox® dry film lubricants. The method of application is to apply the fluid to the object to be coated, then after the carrier fluid has evaporated, to heat the object to the fusion temperature of the Teflon microparticles (~300°C to 400°C) for five to ten minutes in order to fuse the microparticles into a continuous thin film. This thermal melt step is definitely necessary without thermal fusing the Krytox coating is porous and can be removed by light wiping. When using the Krytox line of fluids, the Teflon is optimally applied after the welding and trimming operation before the piezoelectric element is attached. This is because the temperatures required to post process the Teflon film is sufficient to damage the epoxy bond and depolarize the piezoelectric driver. When applying the Teflon films to finished microdrop generators, we internally pressurize the drop generator, then dip coat the tip of the aperture into the Teflon carrier fluid. The fluid applied to the drop generator is then allowed to evaporate with the pressurization being constantly maintained. The particles in the Krytox fluids are specified as being up to 35 microns in diameter. The pressurized air coming out of the ejector hole keeps this passage from being blocked. The drop generators are kept pressurized during the heating to keep the melted Teflon particles from closing off the hole. Maintaining pressurization while placing the drop generators in a temperature controlled curing oven may be difficult. The method that we have adopted to do the thermal melt is to use an electronic assembly heat gun rated for at least 400°C and direct the hot air jet at the tip of the drop generator tube while the rear of the tube is connected to a source of pressurized air. This should be done in a well-ventilated area since the fusion temperature also is within the temperature range that starts to produce toxic fluorocarbon decomposition products. The Krytox coating will undergo a distinct phase change, turning translucent when the melt temperature is reached. This is what we used as an effective gauge of temperature. When cooled, a properly process coating should be smooth to the touch, as opposed to the powdery finish of the originally applied Krytox. The Krytox film is usually several tens of microns thick when finished. Durability is good. The film can be remelted if it is scratched or damaged. The piezoelectric element can be protected by shielding it with metal foil and using water soaked cloth to keep the temperature lower than the material's Curie point. One problem that we have noted in this coating is that the surface finish is relatively thick and can be rough and irregular on the order of microns in feature height. This can cause significant deflection of the drop ejection direction. For certain scientific measurement applications, this may not be important. If the drop generator is to be used for application requiring deposition of materials to exact locations the deflections produced by the rough finish may adversely affect its use for these purposes.

A different type of fluoropolymer high contact angle thin film coating fluid is manufactured by Cytonix Corporation (8000 Virginia Manor Road, Beltsville, MD). These go by the trade name FluoroPel® solutions. Unlike the Dupont Krytox fluid, which carries micron-sized solids, the Cytonix FluoroPel fluids are true solutions. One consequence is that the surface is smooth, gauged by interference film effects, to submicron dimensions. These solution-based fluoropolymer coatings, as applied by dipping, result in much thinner, smoother coats than fluids based on carrying suspended solids. Cytonix manufactures its fluoropolymer coatings with carrier fluids with different characteristics for adhesion to different substrates and setting by evaporation or by catalysts or UV light. We used the PFC 801A, PFC 801A/coFS, PFC 1601A and PFC 1600V fluids successfully on our ejection apertures. The curing temperature is ~100°C, though these fluids have sufficient adhesion and uncured surface smoothness to function well even without thermal curing. Contact angle for water is claimed to be from 120° to 170° for water (Table 9.2). The thickness of these films is on the order of a few microns. Unlike the Krytox fluids, the Cytonix fluids do not appear to liquefy under high temperature curing. This lack of liquifi-

Table 9.2   Measured Contact Angles for
10 Microliter Water Drops on
Different Surfaces

Substrate	Contact Angle
Glass	15°
Bare silicon wafer	15°
Silicon nitride thin film	30°–40°
Silane thin film	65°
Cytonix 801A	110°
Cytonix 801A/FS	110°
Cytonix 1601	120°
Teflon (plasma) thin film	110°

Table 9.3   Measured Contact Angles for 10 Microliter Dow Corning 5 cS Silicone Oil Drops on Different Surfaces

Substrate	Contact Angle
Glass	< 5° constantly spreading thin film
Bare silicon wafer	< 5° constantly spreading thin film
Silicon nitride thin film	< 5° constantly spreading thin film
Cytonix 801A	35°
Cytonix 801A/FS	35°
Cytonix 1601	45°

Table 9.4   Measure Contact Angles for 2.5 Microliter Fluid Drops on Different Surfaces

Fluid	Glass	Silicon	Silicon Nitride	Fluoropolymer Coating
Distilled water	<5°	30°	80°	110°
Propylene glycol	<5°	25°	45°	90°
Ethylene glycol	<5°	25°	45°	90°
Glycerol	10°	30°	60°	105°
1,3 Butanediol	7°	25°	35°	95°
Silicon oil (5 cS)	<5°	<5°	<5°	40°
Ultraol mineral oil	<5°	<5°	15°	75°
Castrol 5W30	<5°	15°	20°	80°
Meteorite suspension	<5°	15°	20°	70°

The contact angles were evaluated by eye with a magnifier and a reference angle chart.
Estimated measurement error ± 20%.
The fluid volumes used was 2.5 micro liters.
The glass surface was a standard untreated microscope slide.
The silicon surface was a standard bare silicon wafer. Since this wafer was exposed to air the actual surface was a thin film of silicon dioxide.
The silicon nitride surface was high stress thin film silicon nitride on a silicon wafer.
The fluoropolymer coating was Cytonix 804A/coFs applied to a microscope slide.
The meteorite suspension was composed of a carrier fluid (95% Ultraol mineral oil, 5% Castrol 10W30 motor oil, [Burmah-Castrol plc., Swindon, UK]) with a 20% by weight addition of pulverized carbonaceous chondrite meteorite.

cation combined with the film thickness eliminates the requirement to internally pressurize the drop ejector during thermal curing. The manufacturers imply for some of their fluids that thermal curing is unnecessary for optimal performance if a drying period of a few days is used. Overall, these solution-based coatings are a lot easier to use for surface coating drop ejector apertures than the solid particle based fluids (Table 9.3 and Table 9.4).

In our experience, the trade off between the use of the Dupont microparticle based fluorocarbon film and the Cytonix solution based fluorocarbon thin films is that while the Cytonix films better preserve the physical contours of the ejection aperture and do not require a high temperature processing step, the Dupont coating has superior ruggedness and has a slightly higher contact angle with most fluids reducing surface wetting to a higher degree. We ended up coating our glass nozzle drop ejectors with the Dupont thin film and our micromachined nozzles with the Cytonix solutions. Since glass nozzles formed by drawing and grinding generally have poor directional ejection characteristics, the use of the Dupont coating with its superior ruggedness and contact angle, although often producing an irregular surface finish, does not significantly alter the nominal ejection consistency of the glass nozzle. The principal advantage of micromachined nozzles, however, is their more consistent fluid jetting performance. In order to avoid compromising the consistent directional jetting characteristics of the micromachined nozzles the fluorocarbon polymer solution based Cytonix fluids were used.

These dip coated fluorocarbon films could be removed by rapidly passing the coated tip of the drop generator through the blue tip flame from a propane torch. The emphasis here is rapidly. Dwell time should be about 1 second in the flame or the tip will deform. The tip should not be pressurized or the glass will blow out. Wiping with an acetone soaked piece of optical cleaning paper also works. This should be done however under conditions of adequate ventilation due to the toxic by-products of fluorocarbon compound combustion.

If aqueous fluids are used, the interior of the drop generator should be kept free of these fluoropolymer high contact angle thin films. While these films may be attractive because they reduce adhesion of contaminants and hence aid cleaning, they seriously interfere with proper operation of the drop ejector. Aqueous fluids no longer wet the surface of the dropper and form a clean meniscus at the ejection aperture hole. We have observed drop generator reservoir tubes internally coated with Teflon where there was a rounded fluid surface internally at the end of the internal fluid column. The fluid was repelled from wetting the walls and preferred to cohere to itself. The inside tip of the drop ejector was dry. Positively applied internal pressure was able to force the fluid to the meniscus, but this was an unstable way to operate since excess pressure would result in the fluid being forced out the tip. With fluids such as silicone oil, which has a low contact angle with Teflonized surfaces, Teflon coated dropper interiors may be less of an operational problem.

The principal gain we have noted was when ejecting fluids with suspended solids. The operating time until cleaning was required was five times longer than when using coated versus uncoated ejection apertures. When ejecting certain nonaqueous fluids such as silicone oils, these coatings had virtually no effect when used without externally applied negative pressure so far as preventing surface wetting. Unlike the case for aqueous fluids, fluid would leak from the aperture hole if negative pressure was not applied. With mineral oil suspensions, the fluoropolymer coatings were found to be necessary for suppressing the gradual fluid buildup over the nozzle hole that would occur even with high levels of negative operating pressure. Neither negative pressure nor anti-wetting coatings alone would suppress fluid build up, but used together reliable operation was achieved.

REFERENCES

1. M.L. Savage et al., A search for fractional charges in native mercury, *Phys. Lett.*, vol. 167B, no. 4, pp. 481–484, 1986.
2. G.L Switzer, A versatile system for stable generation of uniform droplets, *Rev. Sci. Instrum.*, vol. 62, no. 11, pp. 2765–2771, 1991.
3. J.C. Yang et al., A simple piezoelectric droplet generator, *Experiments in Fluids*, vol. 23, pp. 445–447, 1997.
4. R.E. Drews, The effect of translationally symmetric nozzle face defects on the directional accuracy of thermal ink jet arrays, IS&T's Seventh International Congress on Advances in Non-Impact Printing Technologies, Portland, OR, October 6–11, 1991, pp. 107–116, IS&T: The Society for Imaging Science and Technology, 1991.
5. R. Fagerquist, Effect of meniscus formation on the trajectory of a cylindrical jet, IS&T's Seventh International Congress on Advances in Non-Impact Printing Technologies, pp. 67–75, IS&T: The Society for Imaging Science and Technology, 1991.
6. E. Bassous, Nozzles formed in monocrystaline silicon, U.S. Patent 3,921,916, 1975.
7. R.M. Finne and D.L. Klein, A water-amine-complexing agent system for etching silicon, *J. Electrochem. Soc*, vol. 114, pp. 965–970, 1967.
8. H. Seidel et al., Anisotropic etching of crystalline silicon in alkaline solutions l. orientation dependence and behavior of passivation layers, *J. Electrochem. Soc.*, vol. 137, no. 11, pp. 3612–3626, 1990.
9. H. Seidel et al., Anisotropic etching of crystalline silicon in alkaline solutions II. influence of dopants, *J. Electrochem. Soc.*, vol. 137, no. 11, pp. 3626–3632, 1990.
10. O. Tabata et al., Anisotropic etching of silicon in TMAH solutions, *Sensors and Actuators A*, vol. 34, pp. 51–57, 1992.
11. G. Delapierre, Micromachining: A survey of the most commonly used processes, *Sensor and Actuators*, vol. 17, pp. 123–138, 1989.
12. K.E. Petersen, Silicon as a mechanical material, *Proceedings of the IEEE*, vol. 70, no. 5, pp. 420–457, 1982.
13. R.L. Gamblin, Orifice Plate Construction, U.S. Patent 4,528,070, 1985.
14. J.L. Dressler, Two-dimensional, high flow precisely controlled monodisperse drop source, Fluid Jet Associates, Spring Valley, OH, SBIR phase II report, March 15, 1993.
15. S.L. Zoltan, (Clevite Corp.), Pulse droplet ejection system, U.S. Patent 3,683,212, 1972.
16. D.J. Hayes and D.B. Wallace, Overview of small holes, Conference, Non-Traditional Machining, MS89184, Orlando, FL, October 30–November 2, 1989, Society of Manufacturing Engineers, 1989.

CHAPTER 10

Drop Ejector Construction

10.1 TUBULAR RESERVOIR PIEZOELECTRIC DROP EJECTORS

After reviewing the literature, researchers at the Microdrop Particle Search Group at SLAC found that the cheapest to construct and the easiest to service and maintain drop-on-demand ejectors for laboratory use were variations on the piezoelectrically driven tubular reservoir ejector, originally designed and patented by Steven Zoltan of the Clevite Corporation.[1,2] The principal practical advantages over other designs are the ease of fluid handling, the lack of need for complex micromachining steps, and the low cost of the components and drive electronics. Another advantageous factor is that licensing and royalty fees are not a problem for the end user, because the patent on this general design, issued in the early 1970s, has expired.

Tubular reservoir drop ejectors are available commercially at approximately $500 a unit for the ejector head alone. Since drop ejectors are often used in quantity over the duration of a given project due to the need for different diameter microdrops, the need to have multiple units available containing different fluids, and the need to replace damaged units, this commercial cost per unit can become excessively expensive. One may also need to construct drop ejectors that are optimized to one's own experimental requirements. For these reasons, the research group learned to manufacture our own ejectors and developed, over a period of years, a large number of variations on the basic Zoltan design. Some of the more interesting variants with details of their methods of construction are in the next sections.

The Zoltan designed drop ejector uses a tubular fluid reservoir with one end terminating in an ejection nozzle. A piezoelectric element located at one point along the tube provided pressure pulses that cause ejection of fluid through the ejection aperture. The opposite end of the tube from the ejection aperture usually has a fluid feed or pressure control hose attached, which provides controlled negative pressurization to optimize drop ejection. This pressure control hose also can provide the positive and negative pressurization needed to load and purge fluid from the tip. The design features that make this type of microdrop ejector particularly useful for scientific research is the ease of making all fluid contact surfaces out of chemically inert materials, the adaptability of this design to operate with extremely small fluid volumes, and the low cost of fabrication.

This tubular design surprisingly does not require that the fluid is filled to the level of the piezoelectric element location. A very small amount of fluid in the tip, less than a microliter, is sufficient for ejection to take place, even if the piezoelement is located centimeters away from the ejection aperture end. This allows construction using suitable materials, of a microdrop generator with very little dead volume. This can be very important when using "hyperprecious reagents." These are biological reagents that are extremely costly to synthesize or are very scarce. The disadvantage in operating with only a small amount of fluid in the tip is that the drive settings must be constantly changed as fluid is ejected and the level in the reservoir tube is reduced. For instance, there can be up to a factor of five increase in the amplitude needed, as the fluid level decreases from 5 microliters to half a microliter in a 1.5 mm internal diameter (ID) reservoir tube. Fortunately, in our experience, the only change needed has been amplitude. Pulse width retuning has not been necessary.

Another reason for using the partial filling of only the tip is if high temperature fluids, such as liquid metals, are to be ejected. If only the tip needs to be heated, the piezoelectric element and the adhesive bonding agent attaching it to the reservoir tube are spared exposure to potentially damaging high temperatures.

The tubular fluid reservoir both contains the fluid to be ejected and transmits the mechanical impulse from the piezoelement to the fluid. The materials requirement for this tube is that it be chemically compatible with the fluids used and be sufficiently rigid that it does not damp out the drive pulse. Also, it is of great practical advantage if this material is transparent so that the fluid level and the presence of air bubbles can be readily determined.

Glass is an obvious and nearly ideal material for the reservoir tube. Glass tubing is commercially available with internal bore diameters of down to tens of microns, if this is required. Not all types of glass are suitable, however, for mechanical engineering reasons. In particular, the assembly operations needed to form or attach an ejection aperture often require thermal glass forming operations. Different types of glass have radically different thermal forming (glass blowing) behavior.

Common soda lime glass forms at low temperatures (~700°C), which is convenient in that it can be worked with a common propane hand torch. It has rather unfortunate thermal expansion behavior in that unless it is carefully annealed, the heat formed sections will develop stress cracks and come apart. This difficult-to-remove thermal stress makes it very difficult to thermally weld this type of glass to micromachined silicon aperture plates, for instance. What usually happens is that the weld looks good initially, but after it cools, cracks start forming at the weld and gradually turn into a crazed region. This crack formation can take place over a period of time ranging from a few minutes to a few hours. Normal handling after this stage results in the welded aperture plate falling off with the glass in the region of the weld turning to powder.

Pyrex glass tubing did not exhibit this tendency. In fact, glass-to-silicon welds to Pyrex, even without any deliberate annealing, stayed intact without any problems. Pyrex softens at a higher temperature (~900°C) than soda lime glass, but is still within the working range of propane hand torches. Glass-to-glass welds and blown glass sections of tubing were all more reliable with Pyrex due to the absence of thermal stress related problems. Quartz is even better than Pyrex so far as mechanical ruggedness and lack of thermally induced stresses, but has such a high working point (~1800°C) that an oxygen-assisted flame is needed to work it. We have used Pyrex almost exclusively and have found no problems, even without special cool down or annealing steps.

We also have constructed drop ejectors from metal tubing, attaching the ejection apertures to the metal tube using epoxy adhesives. This style of construction has the advantage of extreme ruggedness compared to glass reservoir tube drop ejectors, but has the disadvantage of not having a readily inspectable fluid level. Metal drop ejectors can also chemically react with many fluids one may wish to eject. Visual inspection of the fluid for air bubbles is also not possible. Finally the range of solvents that can be used to clear clogged apertures is far more limited for metal drop ejectors with adhesively attached aperture plates as compared with glass constructed drop ejectors with thermally welded or thermally formed aperture plates.

The fluid ejection aperture hole, for most applications, needs to be in the range of 10- to 100-microns in diameter. The optimal side profile of this aperture hole is that of a cone to avoid the problems of excessively high fluid impedance for a long, straight cylinder and the fragility of a hole punched into a wide, thin, film membrane. If a cylindrical hole is used, it is desirable to avoid having aspect ratios for the hole of more than four to one for the ratio of the hole depth to the hole diameter to minimize the fluid impedance. Long cylindrical ejection holes are also highly vulnerable to clogging and being extremely difficult to clear after solid materials lodge in the nozzle channel.

10.2 METAL BODY DROP GENERATOR

These drop generators were constructed for a particle physics experiment in which the only ejection apertures available at the time were in the form of laser-drilled, stainless steel, 9.5-mm diameter discs.

The body of this drop ejector was turned on a lathe. The internal restrictor nut was an allen head set screw with a 0.4-mm hole drilled through its center. The piezoelectric element and the aperture plate are attached with epoxy adhesive. The nylon washers are present as spacers to hold the piezoelectric element in place while the epoxy sets. The manometer tubing used was Tygon® (Saint-Gobain Performance Plastics, Akron, OH), which was chosen for its transparency and impermeability to air (Figures 10.1 and 10.2).

The design intent of this drop generator was to improve fluid ejection efficiency by enclosing a small volume of fluid between the aperture plate and a fluid restrictor (in the form of a 0.4-mm drilled allen head set screw). The ejection aperture and the piezoelectric disc are attached to the brass reservoir tube with epoxy glue. Miller

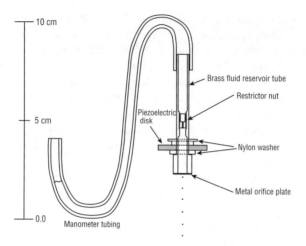

Figure 10.1 Metal body, stainless steel orifice plate microdrop ejector.

Stephenson 907 (Miller-Stephenson Chemical Co., Inc., George Washington Hwy., Danbury, CT) was found to be the most permanent, though Devcon 5-minute epoxy (Devcon, 30 Endicott Street, Danvers, MA) has also been successfully used. There is a very significant difference in the hardness and bond strength between these two types of epoxy. An ejection aperture attached with the 5-minute epoxy could be removed by gently prying with a razor blade. This was seen as advantagous when a permanently clogged ejection hole was found, and it was necessary to change apertures. Removal of a Miller Stephenson 907 epoxy bond, without destroying the glued components, required immersion in methylene chloride (commercial paint stripper) for several days to weaken the epoxy bond prior to mechanical debonding.

After gluing the aperture plate to the body of the drop ejector, we used silver conductive paint to make an electrical contact with the metal reservoir tube. We have found that for many scientific applications, it is desirable to control the charges of the ejected drops. It is necessary to control the potential of the ejection aperture plate for these applications.

The drop generator was operated with the manometer tube, and the fluid reservoir tube continuously filled with fluid. The metal reservoir tube was filled first. Small air bubbles were purged by placing the filled drop generator tube into a vacuum chamber. Small bubbles near the ejection aperture act as fluidic shock absorbers and raised the amount of energy needed to eject drops. The presence of small bubbles also adversely affected drop ejection stability and made tuning difficult and optimal drive setting unstable. The manometer tube was attached and fluid was added to the desired level. Small air bubbles at a distance from the ejection aperture did not seem to have an effect on drop ejection.

The intended ejection fluids were low vapor pressure oils that would be chemically nonreactive with the brass reservoir tube and the epoxy adhesive used to bond the aperture plate to the brass tube. There were definite limitations to the materials' compatibility of this design because metals and adhesives were in contact with the

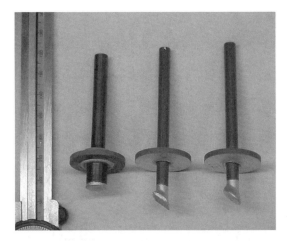

Figure 10.2 Variants on the adhesive attached orifice plate metal body drop ejector. The drop ejectors at the center and the right had the ejection apertures attached at an angle in order to direct the microdrops at an initial trajectory 45 degrees from the axis of the reservoir tube.

fluid. Reactive organic solvents and acids, for instance, could not be used as either ejection fluids or cleaning liquids. In fact, aqueous solutions slowly reacted with and softened the epoxy adhesive. Sterilizing it for use in dispensing biological reagents would have to be done in a gas sterilization setup with toxic gases. In contrast, all glass or thermally welded silicon aperture drop ejectors could be sterilized by immersion of the tip in hot fluids or reactive liquids.

One major advantage that this drop ejector has over the glass and silicon ejectors is a very high degree of mechanical ruggedness. However, for most applications, glass ejectors function well, if handled carefully. The inability to inspect the fluid for fill level and gas bubbles are also serious operational disadvantages compared with glass body drop generators.

The first SLAC-automated Millikan fractional charge search experiment used this type of metal body drop generator for a nine-month run.[3]

10.3 GLASS, STRAIGHT TUBE, WELDED APERTURE

This design for fluid drop ejectors was intended to address some of the problems encountered when utilizing the metal body tubular drop ejector (Figure 10.3). The glass body facilitated inspection of the fluid. It was found that the flow constriction was not necessary for reliable drop ejection. In fact, it was not necessary to have the fluid level above the point where the piezoelement was mounted. A simple tube capped at one end by the ejection aperture with a piezoelectric element glued to the tube could make a functional drop ejector. If the fluid's wetting properties were appropriate, the pressure control manometer could be discarded.

The preferred aperture is constructed from micromachined silicon. This takes advantage of the ability to precisely form holes of a desired diameter and depth

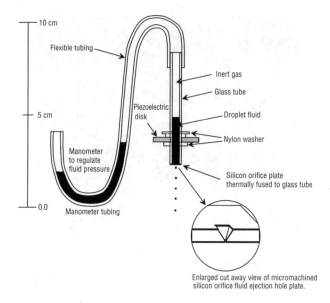

Figure 10.3   Tubular drop ejector constructed from glass thermally welded to a micromachined silicon orifice plate.

using silicon micromachining technology and the ease of thermally welding silicon to glass.

We have also made drop generators by gluing aperture plates onto the glass reservoir tube. This was done for two reasons. One was to attach apertures that had thermally sensitive materials (i.e., Teflon thin film coatings). The other was to attach apertures of materials incompatible with glass thermal bonding.

10.3.1   Fabricating the Apertures

The apertures were fabricated on standard 100 cm × 0.5 mm thick wafers. These wafers were cut into 1 cm by 1 cm squares, each with an ejection aperture in the center. Some micromachining processes leave thin film coatings on the surface of the wafer that may not be optimal for welding. In particular silicon nitride is often used as an etching mask for potassium hydroxide etches. This coating can be removed from the welded region of the wafer by a hot phosphoric acid bath or by sanding with fine grit (600 to 2000) carborundum paper. Silicon nitride can be welded to glass, but the glass does not wet the surface as readily as it does bare silicon.

As an alternative to microfabricated silicon ejection apertures, commercially available sapphire or ruby jewel bearings or nozzle holes of the required diameter can be used. Glass will readily thermally weld to ruby and sapphire. The principal disadvantage is that the aspect ratio (the length-to-diameter ratio) of the holes in off-the-shelf standard jewel bearings is far too high for optimal fluid ejection. The increased fluid impedance of using long aspect ratio orifices can raise the ejection pulse amplitude required by factors of 4 to 10 over that of an optimized microma-chined structure.

10.3.2 Thermal Weld Attachment of the Ejection Aperture

The glass reservoir tube is prepared for welding by squaring off and grinding flat the end to be welded. The opposite end is flame polished to remove sharp edges that may cut shavings off of flexible tubing used to regulate internal pressure. There are two methods that we have used to weld the glass to the silicon.

Glass tubing that is large enough in diameter that a 1 cm × 1 cm or larger silicon aperture chip can be balanced on the end can be welded by supporting the tube from the bottom with a fixture and directing the flame from a propane torch down onto the wafer. The area of the wafer must be sufficient to shield the glass tube from the direct heat of the flame. Otherwise, the glass tube will distort and start to melt. About 1 cm by 1 cm was found to be the lower limit when using conventional propane hand torches. The wafer should be heated until it is glowing orange. The time for a secure weld to form is approximately 6 minutes. The oversized wafer, if desired, can be trimmed flush to the outer surface of the reservoir tube. The principle advantage of this technique is that no alignment or holding hardware is required. The disadvantages are that precise centering of the ejection aperture plate on the glass tube is difficult and that this technique is not practical for welding to small-diameter glass tubing.

Alternately the silicon aperture plate can be placed on a flat, thin piece of a heat-resistant material (such as silicon), which can act as a heating plate when a torch is directed on it from below. This has the advantages of facilitating real time inspection of the progress of the weld joint as heating is in progress and being useful for welding to small-diameter glass reservoir tubes. Small-area silicon chips are also weldable using this method, since the lower heating plate shields the glass tube from the direct heating of the flame (Figure 10.4). The welding assembly rig in Figure 10.5 holds the glass tube in an adjustable x-y stage, which allows precision centering of the tube over the portion of the silicon chip that contains the micromachined ejection aperture. After the glass tube is centered over the ejection hole, the clamp holding the glass is released, and it is dropped onto the silicon chip resting on it, though restrained from tipping by the presence of the loosened clamp. A propane torch flame is then applied to the underside of the support plate and the aperture is welded. The welding time for this style of welding is shorter than that for the method where the aperture is on the top of the glass tube, because the weight of the glass reservoir tube helps join the two pieces together. Applying a vacuum to the inside of the reservoir tube shortens the welding time to less than a minute, but often produces a distorted tube and a slanted weld if the heating is not totally uniform. Allowing gravity to press the surfaces together was found, in practice, to be more forgiving of nonuniform heating and nonaxially symmetric mechanical holding jigs. The progress of the weld can be observed by noting the change in appearance of the contact point between the bottom of the glass tube and the silicon aperture.

10.3.4 Trimming

After welding, the excess silicon is trimmed to aid in cleaning, and to enhance the mechanical ruggedness of the finished structure. Trimming is best done by a

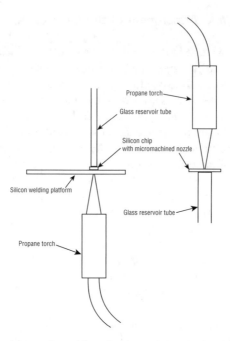

Figure 10.4 Methods of thermally welding ejection nozzles to glass reservoir tubes. Silicon chips larger than 1 cm containing micromachined nozzle structures can be rested on top of the glass reservoir tube and heated to the temperature needed to melt the glass beneath to form the thermal weld. The silicon chip must be large enough to shield the glass from direct exposure to the flame jet. The glass reservoir tube must be large enough in diameter for the silicon chip to rest stably while being heated with a torch flame jet. Small-diameter glass reservoir tubing can be welded to correspondingly small-size silicon chips by resting the silicon chip on a silicon welding platform which acts as a heat transfer plate and flame shield. Large size silicon chips can also be welded in this fashion. The principle advantages of welding on a heat transfer plate are the ability to better align the nozzle structure with the center axis of the glass reservoir tube and the ability to inspect the progress of the glass to silicon weld during the heating operation.

rapidly rotating abrasive wheel (Figure 10.6). Trimming by hand against stationary, fine grit abrasive paper is possible, but runs the risk of cracking the glass tube or fracturing the silicon aperture plate if excessive pressure is used. The major practical problem with sanding by hand is that if light enough sanding pressure is used to preclude damaging the drop ejector, it can take an excessive amount of time. A motorized grinder is far more practical.

A motorized grinder set up under a binocular microscope also allows ready monitoring of the progress of the work while the material removal is in progress. A practical setup for this might consist of a hand grinder (e.g., Dremel MotoTool) fixed in position under a low-power binocular microscope. Carborundum abrasive paper of 320 to 600 grit is the abrasive grade that researchers found most suitable for trimming silicon apertures. Trimming should be done to bring the dimensions of the silicon aperture flush with or just slightly larger in diameter than the diameter

Figure 10.5 Thermally fusing a silicon ejection aperture plate to a glass tube. The material that the glass tube and silicon aperture is resting on is a rectangular piece of 0.5-mm thick silicon wafer. The heat source is a handheld propane torch. The glass tube is initially held out of contact with the ejection orifice plate and is positioned to the desire location above the ejection aperture plate using the micrometer adjusts on the x-y stage. The retaining screws are released and the glass tube rests with its own weight on top of the ejection aperture. The lower silicon rectangular heating surface is brought to a red-orange glow with a handheld propane torch, which maintains temperature until the glass is observed to form a continuous melted bead on the entire contact area with the aperture plate.

of the glass reservoir tube. The direction of the grinding should be to place the glass weld under compressive stress. Glass is very strong in compression but fractures easily under tension at stress risers, such as a thermal weld. The trimming operation typically takes from 2 to 5 minutes. It is advisable to wear nitrile or other puncture-resistant gloves when performing this operation. The spinning abrasive wheel can give a deep cut fairly quickly, if one's fingers touch the edge of the spinning sanding disc. Running one's fingers into the edge of the wheel is all too easy to do, if one's attention is focused on what one is observing through the binocular microscope.

Sometimes while grinding away the excess aperture material, cracks will develop in the glass starting at the weld. This usually is caused by grinding into the glass weld to an excessive depth. In general, try to avoid grinding into the glass at all, if this can be avoided. Grinding a glass weld with a rotating abrasive paper wheel is an invitation to crack production at that point. These cracks, if small and localized to the area near the weld, fortunately can be closed up after grinding is completed by placing the drop generator back into the welding rig and reheating for about 30 seconds with the heating plate at orange hot temperature.

Figure 10.6 Trimming excess material from the welded orifice plate. Cutting away excess aperture plate material eases cleaning and improves mechanical ruggedness. A set up to perform this operation can be constructed from a motorized hand grinder such as a Dremel MotoTool spinning a disc of 400 grit carborundum paper. Precise grinding is facilitated by performing this operation under a binocular inspection microscope.

10.3.5 Attaching the Piezoelectric Element

The piezoelectric element can be attached to the reservoir tube with any space filling adhesive stiff enough to transmit the mechanical impulse from the piezoelectric element to the reservoir tube. We have successfully used quick-setting, 5-minute epoxy, as well as high-strength Miller Stephenson Epoxy 907. There is no apparent difference in the ejection behavior of the drop ejectors when using the low-strength, fast-setting, or the high-strength, overnight epoxies. The fractional charge search group at San Francisco State University reported using high temperature wax to join piezoelectric elements to glass reservoir tubes (Figure 10.7).

The piezoelectric element we used was a 2.5-cm OD diameter 2.5-mm thick disc with a 0.635 center hole. The material is American Piezo Ceramic lead zirconate titanate ceramic APC 855 (APC International Ltd., Duck Run, P.O. Box 180, Mackeyville, PA). This material was selected for its high piezoelectric coefficient. It has a relatively low Curie temperature (195°C), so it is unsuitable for handling molten metals, such as hot solder, unless special construction is used to isolate the ceramic driver element from the heated portion of the fluid reservoir tube. The driver disc is polarized and plated in the thickness mode. We have used both silver and gold plating. We have found that gold plating is significantly more resistant to tarnishing and erosion of the coating in the research environments where we have operated the drop dispensers. In particular, we have had occasion to utilize moderately strong acids, bases, and saline solutions which had seriously deleterious effects on silver plated electrodes. The preferred way of making electrical connections with the piezoelectric drive disc was with a special clip made from a small alligator clip.

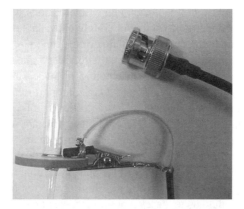

Figure 10.7 Piezoelectric drive disc attached to the fluid reservoir tube using fast setting epoxy. An easily attachable electrical connection to the drive disc can be made as shown in the photo from an alligator clip.

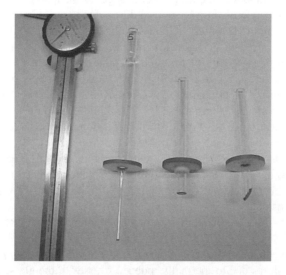

Figure 10.8 Glass body tubular reservoir piezoelectrically driven drop ejectors. The drop ejector at the left is constructed from a Pyrex Pasteur pipette. Standard 6 mm Pyrex glass tubing was used to construct the drop ejectors at center and at the right.

This clip is attached to RG177 small gauge coax terminating in a standard BNC connector. We initially made hard wired solder connections but found that having a permanently attached cable to the drop ejector made initial mounting and alignment, cleaning and storage inconvenient.

10.4 GLASS, PIPETTE, WELDED APERTURE

A straight cylindrical reservoir tube is the most straightforward to construct, but there are applications where it is required that very small volumes of fluid be utilized.

This is particularly applicable for biotechnology applications where reagent and test samples can be so costly per unit volume or simply unavailable in large volumes that it is desirable to minimize the operating volume of the fluid ejector. The obvious way to implement this is to utilize a reservoir tube with a very small inside diameter. We have found practical problems, however, in handling, filling, cleaning and working with very small diameter glass reservoir ejectors. One solution to the problems minimize fluid operating volume while maintaining ease of cleaning and handling in a normal benchtop lab environment, was to use a tapered glass structure with the lower portion near the ejection aperture having a much smaller diameter than the rest of the reservoir tube. A standard glass Pasteur pipette proved ideal for use as the body of this implementation of a small operating volume fluid ejector (Figure 10.8). The minimum operating volume was less than a microliter of fluid. It was not necessary for the drop ejector to be filled with fluid up to the level of the piezoelectric element. The mechanical pulse propagating down the glass tube transmitted the drive energy to the fluid at the bottom of the drop ejector with high enough efficiency that water like fluids required ejection pulse amplitudes in the tens of volts range for microsecond long pulses.

Filling of this style of drop ejector could be done from the tip of the aperture by applying a vacuum via a pressure control tube located in the rear of the reservoir tube. Filling time needed to load a few microliters is a few seconds to a minute, depending upon the viscosity of the fluid and the diameter of the ejection nozzle hole. These ejectors can correspondingly be pressurized to expel unused fluid. Repeatedly filling and expelling low-viscosity solvents was the way that drop ejectors of this type were cleaned during a series of biofluid ejection tests. Since only the tip contacts the test fluid, this type of flushing, if done with a large multiple of the volume of the test fluid, is adequate to clean the drop ejector.

Not all brands of pipettes are equally suitable for making this type of drop ejector. Different brands have different dimensions of the smaller diameter tapered front and are apparently (from welding tests) made from different types of glass. The brand that welded most cleanly was Fisherbrand® Disposable Pasteur Pipettes product number 13–678–20–20A (Fisher Scientific International, Inc., One Liberty Lane, Hampton, NH). We have tried other brands that have proven less than optimal due to insufficient mechanical strength of the tapered glass section to survive mechanical grinding and failures due to thermally induced stresses in the glass, which resulted in the weld turning to powder after cooling.

The piezoelectric element is attached with epoxy in a manner similar to that of the straight tube drop ejector.

This style of drop generator has the unique capability in that if filled only in the narrow tip region, it is capable of operating in any physical orientation due to the fluid location being constrained by surface tension (Figure 10.9). We have run experiments without external pressure control systems with fluids that normally require externally applied negative pressure by simply orienting the drop generators horizontally.

10.5 GLASS, MINIATURE, WELDED APERTURE

One proposed application for these fluid drop ejectors required that they undergo mechanical translation by a piezoelectric bimorph strip. This required that

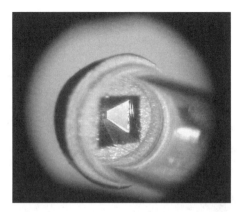

Figure 10.9 Photograph of the tip of a microdrop ejector constructed from a Pasteur pipette. The diameter of the tip is 2.1 mm. The etched pyramidal pit and the ion etched 50-micron diameter cylindrical ejection aperture hole can be clearly seen.

Figure 10.10 Small-diameter reservoir tube drop ejector constructed from 2-mm diameter Pyrex tubing.

the mass be minimized. One solution was to assemble a Pasteur pipette (Figure 10.10) and attach a reduced dimension piezoelectric element to the small-diameter tip section, then cut off the tip section and use just this end of the Pasteur pipette drop ejector. The piezoelectric ceramic element was cut with a diamond-coated steel Mototool cutoff wheel. The hole for the reservoir tube was made with a carbide tip drill. The pulse amplitude required for fluid ejection was higher than that of the larger Pasteur pipette drop generators by approximately the ratio of the area of the piezoelectric elements.

These reduced-area piezoelectric drivers require a higher drive voltage to eject drops. The increase is roughly proportional to the percent reduction of area. For instance, a piezoelectric element with one quarter the surface area would require about four times the pulse amplitude to eject a drop. The impedance is also proportionately reduced so that a pulse transformer can be used to increase the voltage to a level that can drive these reduced area transducers (Figures 10.11 and 10.12).

The resulting waveform from a pulse transformer is not necessarily an amplitude-transformed version of the input pulse. A monopolar rectangular pulse, for example,

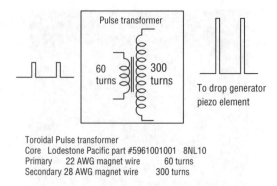

Toroidal Pulse transformer
Core   Lodestone Pacific part #5961001001   8NL10
Primary   22 AWG magnet wire   60 turns
Secondary 28 AWG magnet wire   300 turns

Figure 10.11 Schematic diagram of a pulse transformer for driving piezoelectric drop ejectors.

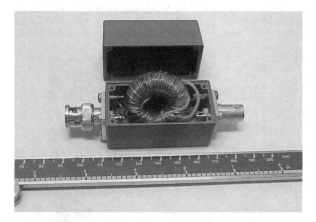

Figure 10.12 Photograph of the pulse transformer diagrammed in Figure 10.11.

will be transformed to a bipolar pulse with equal averaged positive and negative regions. This has relevance to driving drop-on-demand ejectors. It empirically matters which polarity is connected to the piezoelectric elements. For example, for one series of tests ejecting a propylene-glycol-water mixture using the rectangular elements drop ejectors shown in Figure 10.17, one polarity resulted in ejection at a 30% lower amplitude than the other. On the other hand, drop ejection reliability and directional stability were better at the drive polarity that required the higher amplitude.

10.6 PIEZOELECTRIC ELEMENT CONFIGURATION

There are different geometries possible for practical piezoelectric driver elements on tubular reservoir drop ejectors. These different drive element configurations affect the electrical drive amplitudes required, as well as other mechanical engineering issues.

In one experiment, six drop generators were constructed using identical 35-micron diameter ejection apertures and tested for fluid ejection thresholds using a mixture of 75% propylene glycol and 25% water. Single pulse excitation was used with the pulse width varied to find the point of minimum required amplitude. This was usually near 20 microseconds. The drive voltage was measured with an inline oscilloscope. The pulse polarity was switched to find the minimum value orientation. The drop generators also were tested for ejection threshold when driven through a five-to-one step up pulse generator. The optimal pulse widths were different when driven through a pulse transformer. The result of the tests was that for identically sized reservoir tubes, the disc configuration required a slightly lower drive voltage than a cylindrical driver and a significantly lower drive amplitude than the more compact rectangular slab. The smallest drive amplitude design was that of a cylindrical piezoelectric actuator coupled to a minimum length reservoir tube.

The amplitude required to eject drops from the rectangular slab dropper was proportional to the extent of the piezoelectric element. When driven through the pulse transformer, the reduced electrical loading of the smaller, thicker rectangular slab elements bring the pulse amplitude that the drive circuitry must generate to eject drops closer to each other. This is because pulse drive elements have a finite output impedance that limit their peak current. If the piezoelectric element has a lower impedance, then the output pulse can be stepped up more before the increased electrical loading caused by the transformer reduces the effective increase in the voltage transmitted to the piezoelement (Figure 10.13) (Table 10.1).

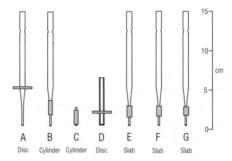

Figure 10.13 Different piezoelectric drive configurations for tubular reservoir drop ejectors.

Table 10.1 Voltage Pulse Amplitude Required for Drop Ejection for Different Piezoelectric Drive Element Geometries (Figure 10.13)

Device	Piezoelement	Ejection Threshold	Ejection Threshold (5:1 Transformer)
A	Disc — 2.5 cm dia by 0.3 cm	18 volts	5 volts
B	Cylinder — 1.6 cm by 0.5 cm	30 volts	9 volts
C	Cylinder — 1.6 cm by 0.5 cm	16 volts	4 volts
D	Disc — 2.5 cm dia by 0.3 cm	42 volts	11 volts
E	Rect — 1.5 by 0.7 by 0.5 cm	74 volts	9 volts
F	Rect — 1.1 by 0.7 by 0.5 cm	122 volts	14 volts
G	Rect — 1.0 by 0.6 by 0.5 cm	115 volts	11 volts

The drive pulse source was a commercial gated voltage source pulse generator. The test fluid used was a mixture of 75% propylene glycol and 25% water.

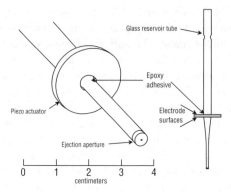

Figure 10.14   Construction details of Pasteur pipette glass body drop ejector.

10.7 PIEZOELECTRIC DISC ACTUATORS

The piezoelectric disc produces fluid ejection by operating in compression, which contracts the volume of the glass fluid reservoir in the region of the hole in the driver disc (Figure 10.14). This produces a pressure pulse that forces out a fluid jet from the nozzle. Fluid ejection surprisingly also takes place, even when the level of the fluid is below that of the location of the driver disc. A cylindrical acoustic wave propagating down the reservoir tube is probably responsible for coupling the pulse energy to the fluid near the nozzle. The main mechanism for producing fluid ejection when the level of the fluid is below that of the piezoelectric element is probably different as evidenced by the large increase in drive amplitude required, as well as the nonmonotonic changes in pulse energy as a function of fluid height.

The practical advantages of the disc-type drive element are low-drive amplitude, low-cost of the piezoelectric element, ease of access to the electrode surfaces, and ease of assembly. The price of the plated disc elements in quantity is about $10 each. This is in contrast with the cylindrical drive elements which cost about three times more. The major disadvantage of the piezoelectric disc drive configuration is that it is difficult to assemble into a spatially dense array. For laboratory use where dense arraying of drop ejectors is not an issue, the disc configuration has been our favored design.

The pulse amplitude needed to produce drop ejection is inversely proportional to the area of the disc. If one's fluid properties are favorable, then the dimensions of the disc can be reduced with a compensating increase in the drive voltage. If one's drive electronics are stable with an inductive load, a pulse transformer can be used to increase the amplitude of the drive voltage pulse. Since the electrical loading of the output stage of the drive electronics is proportional to the area of the piezo-electric element, in principle, one can use a pulse transformer and reduce the area of the disc without needing to increase peak voltage output from the drive electronics.

It is necessary that there be an intimate mechanical bond between the piezoelectric element and the glass fluid reservoir tube (Figure 10.15). In the ejectors that researchers designed at SLAC the piezoelectric drive disc is attached to the glass

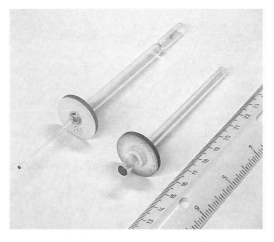

Figure 10.15   Photograph of glass body drop ejectors diagrammed in Figure 10.14.

reservoir tube using epoxy adhesive. Both high-strength Miller Stephenson 907 and fast-setting Devcon 5-minute epoxy were used with no difference in device performance noted. When assembled onto a tapered glass tube, such a Pasteur pipette, the diameter of the center hole should come to rest with the piezoelement placed in the desired location. The glue can be introduced into the gap between the glass and the piezoceremic element and the piezoelement rotated to distribute the glue and remove trapped air. When assembling onto nontapered, straight tube reservoirs, we have used O-rings, friction fitted nylon, or Teflon washers to hold the piezoelectric element in place while the glue hardened. A research group at San Francisco State University reported assembling tubular microdrop ejectors in which high melting-point wax was used to attach their piezoelectric elements to their fluid reservoir tubes. The advantage of this method is ease of removal of the piezoelectric element for reuse, should the glass tube be damaged.

10.8 RECTANGULAR SLAB DRIVE ELEMENTS

This configuration (Figure 10.16) was experimented with in an attempt to find a compact drive geometry for the piezoelectric element. This geometry causes the piezoelectric element to directly compress the fluid reservoir to induce a pressure pulse in the fluid. The advantages of this configuration are compactness, ease of access to the electrodes, and low cost of fabrication. The major disadvantage was the high-voltage amplitudes required to produce drop ejection.

The piezoelectric element was attached to the glass reservoir tube using Devcon fast-setting epoxy. The assembly operation started with a pipette assembly, which had an ejection aperture welded to the end of the glass tube and ground flush. Freshly mixed epoxy is smeared over the portion of the glass tube that the piezoelectric element is intended to be located. The piezoelement is then slid over the region, more epoxy is applied to the ends of the element, and then the piezoelectric element

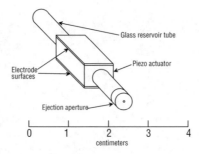

Figure 10.16 Construction details of a direct compression geometry piezoelectrically driven drop ejector.

Figure 10.17 Photographs of direct compression geometry piezoelectrically driven drop ejectors detailed in Figure 10.16.

is worked back and forth and rotated to push out any internal trapped air. Before the epoxy hardens, ethanol soaked Q-tips® (Chesebrough-Pond's Inc., New York, NY) are used to remove the excess glue from the surface of the glass tube (Figure 10.17).

10.9 CYLINDRICAL PIEZOELECTRIC DRIVE ELEMENTS

Cylindrical piezoelectric drive (Figure 10.18) elements were what Steven Zoltan specified in his designs[1,2] for tubular body microdrop ejectors. Our experience indicates that his choice was an excellent optimization of compactness and low-drive voltage requirements. The main disadvantages are higher cost and more difficult assembly over the disc actuator designs.

Assembly options are more complex than with the disc or rectangular slab configurations, due to the problem of how to access the inner electrode surface in the cylindrical drive element. For laboratory use in which the drop ejector may have to be periodically flushed, cleaned, and refilled with different fluids, having perma-

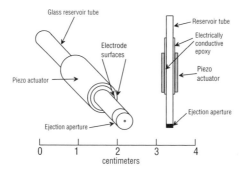

Figure 10.18 Radially polarized cylindrical drive element drop ejector. The piezoelectric drive
element was bonded to the glass reservoir tube using electrically conductive
epoxy glue to facilitate using spring clip electrical connections.

nent wires attached is often inconvenient. One assembly method we used attached
a cylindrical piezoelectric element to the glass tube with electrically conductive
epoxy to bond the piezoelectric element to the fluid reservoir tube as well as to
provide electrical contact. The electrically conductive epoxy was highly viscous and
had to be worked in between the reservoir tube and the internal surface of the
piezoelement using a fine wire. The conductive epoxy was smeared along the length
of the glass to the rear of the piezoelectric element to furnish the electrical contact
point to the inner electrode. External clips were used to attach to the outer electrode
of the cylinder and the conductive epoxy, which acted as the conductive contact to
the inner electrode (Figure 10.19).

As a matter of practical construction technique, it proved impossible to prevent
the conductive epoxy from wetting the ends of the cylinder during assembly, which
if left as is would result in shorted electrodes. To address this problem, we wiped

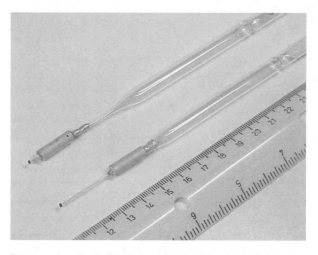

Figure 10.19 Photographs of cylindrical piezoelectric drive element drop ejectors detailed in
Figure 10.18. Conductive epoxy is used to access the inner electrode and bond
the drive element to the glass reservoir tube.

the ends with the sharpened end of a toothpick to thin out the epoxy film, while the epoxy was still drying. Then, after the epoxy hardened, we took 400 grit carborundum abrasive paper and lightly sanded the edges of the ends to break the electrical contact between the inner and outer electrodes.

If the permanent attachment of wires is not an operational inconvenience, they can be solder attached to the piezoelectric element. Soldering wires to the piezoelectric element can be tricky. Too much heat can depole the element by exceeding the Curie temperature of the ceramic material and can thermally debond the conductive coating. To solder contacts, one should use light gauge wires to avoid mechanically stressing the contact point. Enamel insulated magnet wire of 30 gauge or finer or Kynar insulated wire wrap wire is suitable. One caution is that the bond of the wire to the piezoelement is very weak under tension. The conductive coating also is easily pulled off of the piezoelectric ceramic material. A temperature settable soldering iron should be used. Depending upon the solder used, the tip temperature should be set to 400° to 500°C. The best solder to use should have about 3% silver content if silver-based contact materials are to be attached, to minimize leeching of the coating into the solder joint. Kester 44 Rosin flux core SN62 0.015" diameter solder (Kester Solder, 515 Touhy Ave., Des Plaines, IL) is suitable for making these connections. After the connections are made, the wires should be strain reliefed. One practical advantage of using solder wired connections is that it is not necessary to use conductive epoxy to bond the piezoelectric element to the glass reservoir tube. Silver conductive epoxy is highly viscous and is very difficult to work into the thin gap between the glass tube and the piezoelectric element.

A number of compact drop ejectors using cylindrical drivers were constructed to explore the effects of having the majority of the structure of the drop ejector be directly tightly coupled to the piezoelectric actuator (Figures 10.20 and 10.21). This was done to minimize resonant vibrations that would exist in the long reservoir tube

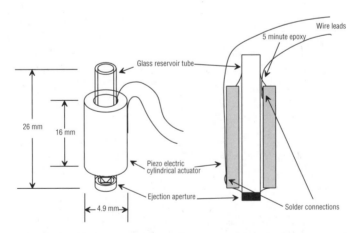

Figure 10.20   Radially polarized cylindrical element piezoelectric drive element drop ejector. For some apparatus, permanently attached drive wires are more suitable for making connection to the piezoelectric element than spring clips. The drop generator diagrammed above had the lowest electrical drive amplitude requirements of any of the drop generators we constructed.

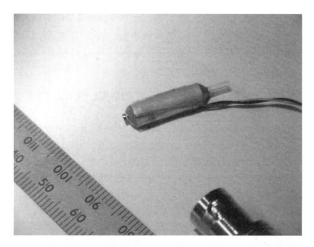

Figure 10.21   Photograph of the cylindrical drive element drop generator diagrammed in Figure 10.20.

designs. Driving the piezoelectric element with a high-bandwidth feedback, controlled, linear, lower amplifier, in principle, provides for a way to actively damp out any unwanted mechanical ringing. There has been some speculation that one major reason for the sometimes narrow pulse excitation range that will eject stable monodisperse drops is that strong mechanical resonances in the drop ejector reservoir tube were being excited, which disrupted the clean breakup of the fluid jet. In one experiment, we performed with 100-micron diameter ejection apertures and distilled water as the working fluid. We were unable to obtain stable, monodisperse drop production with a long tubular reservoir drop ejector, but were able to easily obtain stable operation with the short tube device shown in Figure 10.20. For other more ejection optimized fluids, such as glycol based mixtures, the drop ejector drive geometry had little effect on the ability of the drop ejector to produce reliable monodisperse drops-on-demand.

These short reservoir tube cylindrical driver drop ejectors tend to operate best with relatively longer actuation pulses (10 to 30 microseconds) than the long reservoir tube drop ejector, which usually tuned up best at pulse widths between 1 to 10 microseconds. The short tubular reservoir drop ejector had the lowest ejection threshold of any of the designs that we have constructed. We were able to eject drops of water with as little as a 7-volt drive pulse amplitude.

For use with ejection aperture holes of greater than 25 to 20 microns in diameter, depending upon the viscosity of the fluid, these drop ejectors are optimally filled by attaching a flexible hose to the rear of the reservoir tube and drawing in the working fluid using an applied vacuum. The best tube material for this task is silicone based tubing material, because of its flexibility and low coefficient of sliding friction over glass. Glass is very weak in tension, so tubing that grips the glass too strongly may fracture the glass reservoir tube when working the flexible tubing off the drop ejector.

Short length reservoir tubes with drawn glass ejection apertures, driven by cylindrical piezoelectric elements, are most commonly commercially available.

Companies that sell these devices are listed in Appendix I. The main operational disadvantage of this configuration is the loss of the large integral fluid reservoir that the larger Pasteur pipette based ejectors have.

10.10 INTEGRAL HEATING ELEMENTS

The reliable ejection of some highly viscous fluids may require that the fluid be operated at an elevated temperature in order to reduce its viscosity to a value that allows the ejection of monodisperse microdrops at reasonable drive levels. Integral heaters are relatively easy to incorporate into microdrop ejectors. For some applications, such as the ejection of materials which are solid at room temperature (e.g., liquid solder), provisions must be made to heat the entire fluid reservoir. Construction of the ejector using a high Curie point drive transducer is necessary.

When a heated drop ejector is used for applications in which one is reducing the viscosity of a fluid, which is a fluid as opposed to a solid at room temperature, it is only necessary to heat the fluid at the tip of the ejection nozzle (Figure 10.22). In addition to requiring less operating power, some other advantages of selectively heating only the tip are reduced thermal time constants, which allow more rapid tuning, and refill cycles, and the internal thermal convection currents set up by the tip temperature gradients, which help to keep fluid suspensions mixed.

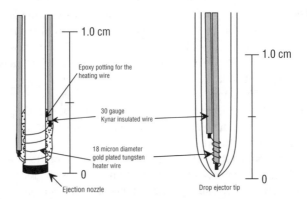

Figure 10.22 Methods of selectively heating the fluid at the ejection nozzle. For reducing the viscosity of a fluid by heating in order to improve its ejectability is not necessary to heat the entire fluid reservoir. Selectively heating only the fluid at the tip is sufficient. We have found that an increase in the fluid temperature of only a few tens of degrees can change a chaotically ejecting fluid into one that will readily produce stable monodisperse drops. Nonelectrically conductive fluids can be heated with minimal applied power with an immersed heating element directly in contact with the liquid. This is not advisable for aqueous fluids due to the potential problem with electrolysis initiated by the voltage difference between the ends of the heating coil.

REFERENCES

1. S.L. Zoltan, (Clevite Corp.), Pulse droplet ejection system, U.S. Patent 3,683,212, 1972.
2. S.L. Zoltan, (Gould, Inc.), Pulse droplet ejection system, U.S. Patent 3,857,049, 1974.
3. N.M. Mar et al., Improved search for elementary particles with fractional electric charge, *Phys. Rev. D.*, vol. 53, no. 11, 1996.

CHAPTER 11

Pressure Control

Most fluids eject much more reliably if negative pressure with respect to the atmosphere is maintained inside the fluid reservoir. This negative pressure suppresses the tendency of the fluid to leak out of the ejection aperture hole and wet the surface of the orifice plate, producing an unpredictably variable fluid layer through which the drop ejector will have to eject its fluid jet. At best this can result in a variable ejection threshold as a function of time and ejection rate. Another negative effect is that the drop ejection direction can be deflected by tens of degrees from its desired perpendicular orientation, if this layer is asymmetric around the ejection aperture by amounts as little as microns. At worst a very thick fluid buildup can totally suppress ejection. An additional failure mode, if the fluid is being used as a volatile carrier for particulates, is that the fluid built up on the outside can evaporate and solidify, thus jamming the aperture.

Negative internal pressure reduces these problems by pulling fluid back into the drop ejector and keeping the external surface of the ejection orifice dry. This is a big plus for reliable operation. Negative internal pressure also tends to reduce the drive voltages needed to eject drops. One can easily find out how much negative pressure a given drop generator fluid combination can take by watching for ingestion of air while slowly increasing the applied negative pressure. This is one type of experiment where transparent fluid reservoir drop ejectors are essential. Ingestion of air by an overpressured manometer does not necessarily stop drop ejection once the negative pressure is removed. These types of air bubbles are large enough that they rise to the top of the fluid volume. The kinds of bubbles that most seriously affect a stable drop-on-demand operation are the very small bubbles that are ingested as a result of improper aperture design or pathological drive settings. These bubbles cling to the walls of the drop ejector near the ejection hole and suppress ejection by absorbing fluid pressure pulses near the region where the fluid jetting takes place.

Most of our experiments have operated with negative pressure of from 4- to 25-cm fluid column of water. Most fluid will operate with a wide range of negative pressures. The principle change to the operating conditions of a drop ejector when the negative pressure is varied through its operating range is that the drive voltage amplitude will tend to decrease as the negative pressure increases. The pulse shape will not require changing, in general. As the ejection aperture diameter increases, the amount of negative pressure that will not result in air ingestion decreases. At the same time, the tendency to leak increases. It is much harder to find optimal negative pressure settings for large aperture drop ejectors (ejection aperture diameter > 75 microns) than for small aperture drop ejectors (ejection aperture diameter < 35 microns).

For most aqueous fluids, the range of operating pressures that will work is increased if a high-contact angle, thin film is increased often to the point where no externally applied negative pressure is needed. Each fluid has to be tested at the required pulse rate to see if this is possible.

11.1 MANOMETER TUBE PRESSURE CONTROL SYSTEM

Most of our experiments with microdrop ejectors have utilized variations on the setup in Figure 11.1. This is simply a modified manometer used to pressure control the volume of air to the rear of the drop ejector reservoir.

We construct the fluid pressure control manometers out of lengths of clear plastic tubing preferably filled with a nonevaporating fluid (Figure 11.1). We normally use low-vapor pressure silicon oil for long duration experiments. For short duration

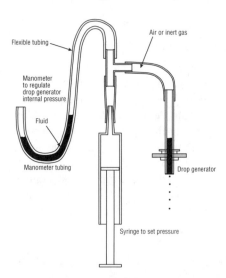

Figure 11.1 Simple manometer-based drop ejector pressure regulator. A difference in manometer fluid height is set by an external syringe and is coupled to the drop ejector fluid reservoir by intervening air or inert gas.

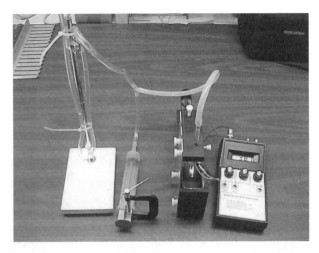

Figure 11.2 Photograph of a simple implementation of a drop ejector fluid pressure regulator. A C-clamp is applied to the plunger of the syringe to prevent it from being slowly drawn down by the negative internal pressure.

experiments in areas where cleanup of spilled silicone oil may be a problem, we have used water. A tee fitting to a large volume syringe with a locking clamp is used to set the negative pressure level. The pressure is read by measuring the relative height of the fluid in the "U" section of the clear tubing. A column of air couples the manometer to the drop ejector. The principal drawbacks of this simple system is that slow leaks seem to be impossible to avoid, which requires resetting the fluid level every few days, and because of the trapped volume of air between the fluid column and the drop ejector, the applied negative pressure will change as the temperature of the room changes. This will shift the optimal pulse amplitude for reliable drop ejection (Figure 11.2).

The necessary fluid height varies with the fluid and the ejection aperture. Most ejector fluid combinations only require between 5 and 20 cm of water density fluid column height to achieve proper operation. Some difficult to eject fluids that have a high tendency to wet the ejection aperture require a greater column height. The design of the U-shaped tube can be as simple as a folded section of tubing wire tied together on a lab stand or, as in Figure 11.3, a floor standing U-tube with a built-in height scale.

11.2 VARIABLE HEIGHT RESERVOIR PRESSURE CONTROL SYSTEM

One practical issue with the pressure regulation systems utilizing differential fluid height in a U shaped section of small bore tubing is that slow air leaks from the multiple connections in the system and requires daily adjustments of the relative fluid heights. One solution to this short-term pressure drift is to design the system to minimize the number of fluidic-air interconnects and reconfigure the fluid geometry such that a small leak will require a much longer period of time to effect a

Figure 11.3 An extended height manometer tube pressure regulator tube.

change in hydraulic head height. The pressure regulation system in Figure 11.4, which utilizes variable height, 100-mm diameter fluid chambers to set the hydrostatic head, was able to produce stable fluid heads over a period of months.

11.3 AUTOMATED PRESSURE CONTROL

For long-term pressure regulation of an experiment intended to run continuously for a year, a system was designed using pressure sensors feeding back to a bidirectional, remotely operated, Masterflex® peristaltic pump (Cole-Parmer Instrument Co., 625 East Bunker Court, Vernon Hills, IL) placed in the pressure control systems to maintain constant negative pressure.

The system in Figure 11.5 and Figure 11.6 utilizes a Lucas NovaSensor® NPC-1210 micromachined solid-state pressure sensor (Lucas NovaSensor, 1055 Mission Court, Fremont, CA) to provide feedback to a computer-controlled pressure regulator. In response to changes in measured pressure, the computer controls a Masterflex® peristaltic pump modified for remotely actuated bidirectional operation to restore the desired pressure level.

In practice, this feedback system was able to regulate internal pressure to ± 0.2 cm column of oil (specific gravity 0.913). In the absence of active feedback, the air dead volume in the system should be reduced to a bare minimum. Normal day-to-night temperature cycling can cause quite significant pressure changes due to the changes in the dead air volumes. An alternate method of setting negative pressure is to totally fill the plastic tube with fluid and set the level of the end of the tube at different

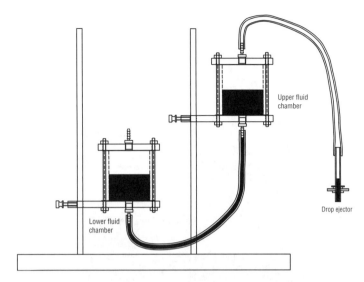

Figure 11.4 Pressure regulator utilizing variable height fluid chambers. The negative operating pressure is set by varying the relative height of two fluidically coupled air tight fluid chambers. One advantage of this technique over the tubing-based systems shown in Figures 11.1 to 11.3 is that slow air leaks take a much longer time to change the hydrostatic head.

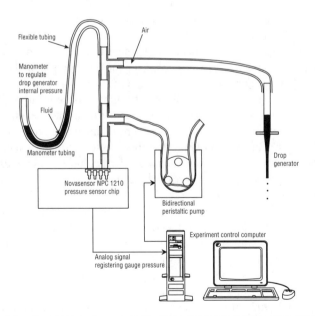

Figure 11.5 A functional block diagram of the computer regulated pressure control system for use in a long-duration, constantly running experiment using microdrop ejection of silicone oil.

Figure 11.6 Photograph of the system shown in Figure 11.5. From the left are the pressure sensor and computer interface chassis, the peristaltic pump, and the manometer fluid column U tube.

heights. This was the method used for the first SLAC automated Millikan experiment. The option was open to use this method because the fluid (silicone oil) was inexpensive, nontoxic, and available in large quantities. It also was believed at that time (mistakenly) to be nonreactive to all of the materials in contact with it. Most fluids that are scientifically interesting do not have these characteristics. This was the reason for the subsequent design for the negative pressure to be delivered, buffered by air or an inert gas. The manometers in the accompanying photographs all deliver negative pressure via an intervening air volume.

The amount of negative pressure that can be applied depends upon the diameter of the ejection aperture hole, the surface tension of the fluid, and the fluid-to-ejection-aperture-surface-wetting properties. A 100-micron diameter aperture might tolerate as little as 4-cm of water fluid column negative pressure. Small diameter apertures of around 10 microns have been operated with fluid columns as high as a meter of water. Typical negative operating pressures can be generated by a fluid height difference of 15 to 30 cm of water.

We have observed custom-designed, industrial, drop-on-demand fluid drop deposition hardware that had fully automated control over the internal pressurization of the drop ejector's fluid in both positive and negative directions. This allowed automated filling, cleaning, and dispensing. There are multiple ways of implementing this type of system, including multiple pressure reservoirs accessed with electrically operated valves or the use of computer interfaced peristaltic pumps. We have constructed a fluid filling and cleaning station with a bidirectional peristaltic pump and manually operated valves for controllably pressurizing and applying vacuum to drop ejectors.

Hewlett Packard (HP, 3000 Hanover Street, Palo Alto, CA) has patented an ingenious way of applying negative internal pressure to its modular ink reservoir ejector units. The HP ink cartridge contains a flexible bladder filled with ink. A porous foam matrix, which is built into the cartridge, tries constantly to expand the bladder to a greater volume.

There is one way of producing negative pressure without using a manometer tube — operate the drop generator inverted with the drops ejecting upwards! Pipette-style drop generators made with small-diameter internal bores are capable of operating in any physical orientation due to the narrow interior fluid channel containing the fluid by surface tension.

CHAPTER **12**

Fluid Engineering for Microdrop Ejectors

12.1 PROBLEMATIC FLUIDS

Not all fluids are ejectable as monodisperse drops from drop-on-demand micro-drop generators, and of those that are ejectable, not all can be ejected with a high enough degree of reliability to be used in a set-and-forget, autonomous, operational mode. The fact that for reliable operation, the working fluid for microdrop ejectors must be a very painstakingly engineered liquid is an unpleasant truth that one does not often find mentioned in marketing literature for commercial microdrop dispensing hardware.

The minimum requirements for an ejectable fluid to be ejectable in any form are:

- It must not contain particulates that are large enough to jam the fluid ejection hole.
- It must not be so viscous that it cannot be jetted.

To form monodisperse drops, the viscosity, surface tension, and volumetric homogeneity of the fluid must be within certain limits. To form reliable monodisperse drops on demand, the added requirements are that the fluid's properties:

- Must be stable with time
- Must be stable under the full range of environmental conditions in which one will operate
- Must not change due to exposure to air
- Must not build up solid deposits in or over the ejection hole
- Must be compatible with the drop generator's materials and ejection process

Examples of some common problematic fluids might be:

- Air hardening adhesives — will solidify in the ejection hole
- Paint — most paints exhibit pigment settling, ordinary paints may clog aperture with pigment particles in addition to hardening in the ejection hole
- Tap water — calcium ions present in most tap water will react with absorbed atmospheric carbon dioxide to form scale (solidified calcium carbonate), clogging the ejection hole
- Motor oil — most are too viscous to be jetted
- Corrosive fluids that react with the drop ejector structure
- Unfiltered seawater — particulates and microorganisms may jam the ejection aperture, calcium carbonate deposits can plate over the ejection hole
- Shear thickening non-Newtonian fluids — too high an effective viscosity when jetted

Conversely, there are some very simple pure fluids and simple fluid mixtures that we have found very reliable. These may be suitable for testing drop ejectors and drop tracking systems or for experiments in which the material making up the microdrop is unimportant such as the use of microdrops to visualize gas flow.

- Mixtures of propylene glycol (or ethylene glycol) and distilled water. A particle filtered mixture of 75% propylene glycol and 25% distilled water is a test fluid that has generated stable monodisperse microdrops in every drop ejector design we have fabricated.
- Dow Corning silicone oils of up to 10 cS viscosity.

Most scientific microdrop applications utilize the ejected microdrops to deliver a substance to a particular location such as a substrate, to an analytic chamber, or to be mixed with other microdrops on a reaction vessel. The general principle of design for a jettable fluid is to find a number of suitable solvents or carrier fluids for the payload one wishes to transport in a microdrop. It is advantageous to have a number of miscible carrier fluids because they allow one to vary the viscosity and surface tension of the final mixture by using different proportions of different carrier liquids. The term "carrier fluid" is used because some payloads must be transported as suspended solids. Depending upon the stringency of the operating conditions, additional fluid components are added to increase operational reliability.

A survey through scientific journals and patent literature very strongly implies that the design of fluids for microdrop ejection is very much an art with a lot of trial and error testing involved to optimize the final mixture. One journal article[1] in a series of three detailing inkjet ink work performed by Xerox researchers[1-3] mentioned that over 2000 inkjet ink formulations were tested in order to develop the four-color ink set for their color inkjet printers!

12.2 NONCLOGGING FLUIDS

If the fluid that one is using clogs the ejection aperture hole, then fluid jetting to form microdrops cannot take place. There are many ways that improperly engineered fluids can clog an ejection aperture hole. The most common reason an otherwise reliable fluid would cause a drop ejector to fail to operate is if a dust particle suspended in the fluid lodges in or over the ejection nozzle hole. Filtering of the fluids, and assembing, filling, and operating in a particle-controlled environment are the best ways to prevent this form of failure. Some drop ejectors incorporate a final particle-filtering stage near the ejection chamber. The particle handling abilities of these final stage filters are limited, however, and cannot compensate for a fluid with a high concentration of large particulates. Another way of producing a particle jam is for agglomerations of smaller particles to occur in an improperly engineered fluid particle suspension. Van der Waals forces will tend to make small particles that come into contact to stick together, forming large agglomerations that can lodge over ejector holes and block off fluid flow. Another mode of particle jamming is for small particles to wedge together in a cylindrical ejection hole and block off fluid flow.

Particle-free fluids can also clog aperture holes. An obvious example of a fluid guaranteed to jam up a microdrop ejector would be an air-drying adhesive that leaves a solid residue after its solvent evaporates. Interactions with the air that can jam ejection apertures can be more subtle. When aqueous fluids containing calcium ions are exposed to air, they acidify due to atmospheric carbon dioxide turning into carbonic acid. The carbonate ions produced by this reaction can react with the calcium and precipitate out as solid calcium carbonate inside and over the ejection aperture hole. It is possible for solid precipitates to form near the ejection hole in an otherwise stable solution due to the concentration gradients that can be created by solvent evaporation within the fluid in the ejection aperture hole.

Exotic materials may clog ejection aperture holes due to their unique properties. Ejection fluids that contain live microorganisms may have the cells form colonies and fibrous masses over time, which can grow over the ejection hole. Molten metal alloys such as hot solder can form clogs by the separation out of the multicomponent metals making up the alloy and having oxides of some of these trace metals form solids on reaction with the air that can then block off the ejection aperture.

Poor fluids can jam drop ejectors both instantaneously, as in the case of a large particle lodging in the ejection hole, or gradually over time as in the case of some aqueous suspensions of ground minerals. In one series of experiments that we conducted with aqueous ground mineral suspensions, the drops became smaller over time and required a continuously increasing drive amplitude to maintain ejection. What we believed was happening was that during each ejection cycle some small amounts of ground minerals and fluid were deposited on the exterior of the drop ejection aperture plate. As the water evaporated, a layer of a cement-like solid was building up over the hole, which eventually choked off the ejection nozzle.

12.3 FIRST DROP RELIABILITY

The first drop problem in inkjet terminology is the difficulty that one often encounters when attempting to eject a fluid drop after a period of latency during which no drop ejection takes place. The basic causes are a milder form of what generates clogs and jammed apertures. Fluids that sit dormant in the ejection aperture hole can experience evaporation, which can result in an increase in relative dissolved solute concentration, which can cause a local increase in fluid viscosity. An increase in fluid viscosity must be compensated for by an increase in the drive energy in order to eject a fluid drop.

A more complex problem can arise if the fluid wets the ejection aperture such that a thin film fluid meniscus forms during operation over the ejection hole. Here, by capillary action, a small amount of fluid will continually leak from the ejection aperture hole and spread across the ejection aperture surface. If the fluid layer is sufficiently thin, the fluid jet that forms the microdrops can fire through this film. When the drop ejector is firing, the fluid in the film gets mixed and cycled by the jet ejection and withdrawal process. As the fluid layer becomes thicker, more energy is required to fire a jet through this fluid. Also, the optimal pulse wave shape to excite the jet may change. As an example, the Microdrop Particle Search Group has used silicon oil as a working fluid for certain long-duration experiments requiring continuous drop generation. Silicon oil has an extremely low-contact angle with most materials and forms an external fluid meniscus even with the fluid reservoir under negative pressure. A steady state can be reached where stable ejection through the thin film meniscus can be sustained for months. However, a cold restart was required hours before the meniscus restabilized, with each major readjustment in drive requiring a similar reequilibration time.

A more serious reliability issue can result from an external thin film meniscus formation from a fluid that is a carrier for suspended solids or a solidifying polymer. It is possible here for a thin crust to form over time that will require a higher than normal drive pulse to force a fluid jet through the aperture.

In practice, the first drop problem is minimized by both fluid and hardware engineering. Humectants are added to fluid to retard evaporation and to facilitate rapid rehydration of the fluids near the ejection hole. High-contact angle external aperture surfaces are used to suppress surface meniscus build up. Piezoelectric actuator drive cycles are used to pull the fluid into the ejection structure, leave a dry hole, and then eject it. Cleaning stations with wiping pads and purge cycles, in which fluid can be ejected at higher than normal pressure into a waste area after a period of latency prior to a new print job, are common in commercial inkjet printers.

In summary, the first drop problem is an operational issue with any drop-on-demand ejector that uses a fluid that has a significant difference in physical state during ejecting and quiescent conditions. For inkjet printers intended for consumer use, the first drop problem must be dealt with or the printer will be operationally useless. For some scientific applications where hand tuning and a long restart time can be tolerated, this phenomenon may be annoying but will not otherwise compromise the use of the microdrop ejector.

12.4 JETTING STABILITY

One characteristic of well-engineered microdrop ejection fluids is that the fluid jet produced and the microdrops formed by each actuation impulse are very nearly identical. However, there are some fluids that experimentally do not have this characteristic. For these difficult-to-stabilize fluids, each actuation impulse produces drops with varying initial velocities, direction of travel, size, and number of accompanying satellites. Sometimes whether a drop is produced at all is a matter of chance. We have experimented with ejecting fluids that have had these characteristics. There were two types of fluid in our experience that produced this type of behavior.

One class of fluid was heterogeneous on a volume scale on the order of the sizes of the drops we were ejecting. Fluids carrying living microorganisms and fluids containing suspensions of crushed undifferentiated minerals were examples of this type of fluid. What we believe is happening is that the fluids making up the fluid jet, which normally destabilizes by the Rayleigh-Taylor process, instead has randomly occurring mechanical discontinuities in random locations on the order of the dimensions of the diameter of the fluid jet. This forces drop break off to be at different points in the jet. The presence of these random, large, suspended solids in the ejection hole also can suppress ejection if they are stacked with other solids in a way that blocks the aperture hole. The negative pressure portion of the impulse can then rearrange them to clear the ejection hole for the next ejection cycle. The directional instability is hypothesized as being due to the presence of a large, nonaxially symmetric, fluid mass in the ejection hole caused by the presence of the random solids.

Other fluids that exhibited irregular drop ejection were certain high viscosity fluids with an extremely low surface tension that tended to form long ligaments between the drop and the ejection aperture. These ligaments would fracture and break up into a random pattern of satellite drops. The pattern of accompanying satellites were different for each ejection cycle. The drops also would exhibit slight jitter in both direction of ejection and velocity.

Unstable drop ejection due to unsuitable values of viscosity and surface tension can sometimes be cured by special additives or different ratios of primary solvents. Another possible solution to the problem of unsuitable viscosity is to alter the temperature of the fluid until the viscosity allows for stable drop formation. We have successfully used drop ejectors with heating elements in the tip to reduce the viscosity of a problematic working fluid until it no longer formed long ejection ligaments (Section 5.7). The problem of large solids randomly destabilizing the fluid jets is best dealt with by increasing the drop size or eliminating, if possible, the largest of the solid particles. Sometimes these are not doable options, for instance, if the fluid with poor rheological characteristics must be jetted in its pure form for analytic reasons. When forced to use intrinsically poorly performing fluid, we have noticed some improvements by using more sophisticated drive waveforms. These waveforms included multiple pulses to excite the ejection of one drop and shaping the rise and fall times of the drive pulses along with their widths. The theory is to excite strong enough resonances in the fluid jet so that drop breakup becomes less of a function of the mechanical discontinuities in the jet and more a function of the externally induced vibrations in the jet. Unfortunately, we have found no way of locating these favorable drive settings other than through a lot of trial and error with each fluid and each drop generator requiring individual tuning. Also, these favorable drive conditions were not stable over time.

Good performing fluids such as aqueous mixtures of anhydrous alcohols, like propylene glycol, eject stable drops on demand with almost any type of excitation pulse over a very wide amplitude range. In contrast, fluids with poor rheologies for ejection require extensive trial and error to find drive conditions that can produce anything resembling monodisperse drops. We have been given some fluids to test that we have never been able to eject as stable monodisperse drops. Unfortunately, one cannot assume that a randomly chosen fluid can be ejectable as directionally stable monodisperse drops.

12.5 DIRECTIONAL STABILITY

Fluid mixtures that produce asymmetric deposits of fluids or solids around the ejection aperture hole can produce serious angular deflections in the direction of the ejected fluid jets. As little as a few microns difference in the height of material around the edge of the ejection hole can produce deflections on the order of tens of degrees.[4] The best solution to this potential problem is to engineer the ejection nozzle such that the outside surface is anti-wetting with respect to the inner surfaces of the drop ejector. In this way, when the fluid undergoes its jetting ejection withdrawal cycle, the liquid jet will tend to pull all fluid residue back into the ejection hole. This is no guarantee, however, that under all operating conditions, some gradual build up of material will not occur on the exterior surface. For instance, if a random spray or small satellite drops are formed with the primary drop, under certain circumstances, they can be pulled back to the surface of the ejection aperture by electrostatic forces and gradually build up surface deposits. Experiments have shown that increasing the velocity of the ejected jet decreases its angular deflection. We have observed dramatic reductions in angular deflection by increasing the drive amplitude. In extreme cases, jets that were deflected at low ejection speeds to trajectories nearly parallel to the surface of the aperture surface straightened out and jetted in a nearly normal angle.

For some reason we do not fully understand, some pulse widths produced less deflected jets than others. One possible explanation is that the vibrational modes excited in the fluid in or over the ejection hole can include nonaxially symmetric modes that produce lateral asymmetry of the fluid layer thickness through which the jet must fire. We have observed a similar phenomenon in high ejection rate operation (kiloHertz) in which small changes in frequencies would cause changes in droplet size, drive requirements for ejection and directional stability. One possible reason for this sensitivity to frequency at a high rate of operation is that the capillary action refill of the nozzle is not fully stabilized prior to the next firing cycle. Certain frequencies of operation would fire the drop ejector with the fluid meniscus in different phases of its final damped fluid oscillations during its refill cycle.

For certain scientific applications where hand maintenance of the drop generator is tolerable, frequent wiping of the surface can minimize the jet deflection problem. We have found that nozzle design can affect the directional stability of the fluid jet. In our experiments, we found that nozzles that terminate with a short cylindrical hole have better directional jet stability than ejection nozzles that terminate in a sharp conical knife edge.

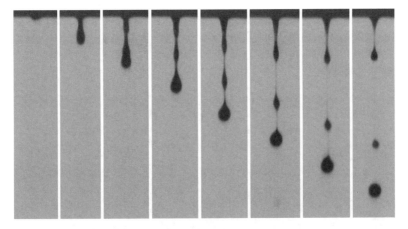

Figure 12.1 Production of repeatable satellite drops. Increasing the drive amplitude of drop ejectors past the point where monodisperse ejection occurs will usually result in an extension of the jet length resulting first in an increase in the drop diameter and then in the production of secondary or satellite drops. For moderate overdrive levels with fluids that have an appropriate surface tension, the satellite formation process is repeatable from one drive pulse to the next.

12.6 FREEDOM FROM SATELLITE DROPS

Many applications where microdrops are used function optimally if monodisperse drops are generated. Some fluids, however, seem to be incapable of reliable generation of a single drop per drive impulse. Instead multiple drops of varying sizes are produced (Figure 12.1).

There are at least three fundamentally different conditions in which satellite drop generation takes place. One condition occurs when overdriving of an otherwise well-engineered fluid past the point where single, monodisperse drops are formed. In the resultant high-speed, fluid ejection event, a longer-than-normal fluid jet is ejected. Its entire length cannot all be drawn into the primary drop or be pulled back into the drop ejector. Under these conditions, multiple drops will form along the length of the jet. This is a highly repeatable process. The multiple drop pattern will repeat exactly, given identical excitation of the drop generator. In well-engineered fluids, this pattern of multiple drop generation in overdriven jets is repeatable enough that at least one physics experiment looking for fractionally charged matter used satellite drop formation in an overdriven jet to generate the small drops it needed, discarding the large primary drops.

Two other satellite producing conditions are due to problematic fluids and will tend to produce satellites regardless of the actuation impulse amplitude. One of these fluid conditions conducive to satellite formation occurs when the fluid is heterogeneous on a spatial scale on the order of the sizes of the fluid jet diameter. This can be due to the presence of large, suspended solids or from the fluid being a too coarsely dispersed emulsion. The presence of inhomogeneities in the fluid jet act as nodal points where the fluid cylinder will preferentially start to destabilize and start

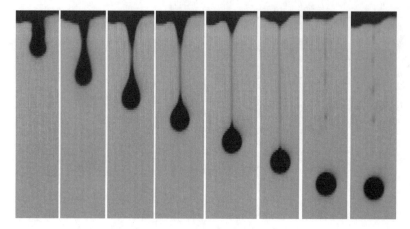

Figure 12.2 Low surface tension fluids can form long, connecting ligaments between the drop and the fluid ejection aperture. Upon breaking up, these ligaments form randomly sized satellite drops.

to condense into spherical drops. Since the location of these inhomogeneities will be random, the pattern of satellite drops will not be the same for each ejection event.

Another type of fluid prone to generating satellites, even when of homogeneous composition, would be a fluid characterized by high viscosity and low surface tension that produces a long connecting ligament between the ejection hole and the primary drop mass at the end of the jet (Figure 12.2). We have observed these very long ligaments to break up chaotically into large numbers of small droplets.

12.7 WIDE JET VELOCITY TOLERANCE

As the excitation impulse to a microdrop ejector using a well engineered fluid is increased, the fluid jet behavior goes sequentially through different phases of behavior as the velocity of the expelled fluid jet increases:

- Expulsion of a fluid jet that is then withdrawn in its entirety back into the ejection aperture
- Expulsion of a fluid jet where a section breaks off to form a single, free drop
- Expulsion of a fluid jet where multiple drops form in a repeatable manner
- Expulsion of a fluid jet that breaks up into a chaotic spray with randomly sized drops

A well-engineered fluid will have a wide jet velocity range where breakup of the jet into a monodisperse drop takes place.[3] This is important for reliable operation of the microdrop generator under varying environmental conditions. For instance, if temperature changes take place, the viscosity of the ejection fluid will change. As the viscosity of a fluid increases, the velocity of the jet will decrease at a given level of excitation. Being able to give a drop generator a healthy overdrive level past the bare threshold at which drop break off occurs, in addition, will help minimize jet deflection caused by debris on the face of the ejection aperture. Fluids with a wide range of drive amplitudes in which monodisperse drops will form can also be overdriven past the drop ejection threshold to compensate for first drop problems.

However, some problematic fluids have either an impractically narrow monodisperse drop production amplitude window or no window all. For some fluids with very poor jetability, all that can be produced is a chaotic spray.

12.8 DRIVE WAVEFORM TOLERANCE

These issues are related to the discussion of having a fluid that has a wide tolerance for different jet velocities, which will result in the formation of monodisperse drops. Piezoelectrically excited drop-on-demand ejectors have a considerable amount of control over the drive pressure impulse used to eject the fluid jet. In addition, the mechanical drive impulse excites different mechanical resonances in the structure of the drop ejector, in the fluid meniscus prior to ejection, and in the fluid jet itself. This drive flexibility has been used in commercial printers to alter the size of the ejected drop to print more effective gray scale halftones.

Some fluids, such as mixtures of propylene glycol and water, will produce monodisperse drops at almost any reasonable ejection pressure waveform. Other fluids will only produce monodisperse drops under a very narrow set of drive conditions. The unfortunate thing about such ejection-condition-sensitive fluids is that as conditions change, such as temperature or humidity or as deposits build up over the ejection hole, that retuning of the drop generator becomes necessary. The operational disadvantages of a fluid that requires continual retuning of the drop ejector are obvious.

12.9 REPETITION RATE

All fluids have a maximum stable drop production rate in a given drop-on-demand ejector. This is governed by two primary factors, the refill rate and the meniscus damping. The refill rate is governed by the time it takes to replace by capillary action the fluid volume lost to the ejection of the previous drop. Fast capillary action refill is facilitated by highly fluidphilic surfaces and a low-viscosity fluid. Fluid oscillations can occur in the fluid meniscus, which must be damped out before a repeatable firing operation can be implemented. If one attempts high-rate operations with fluids that have not had time to damp out its refill cycle meniscus oscillations, one may observe high frequency-dependent operation as the drop ejector fires at different points in the refill process. Fast damping is facilitated by high fluid viscosities, which unfortunately is counter to the low-viscosity optimization for fast fluid flow and rapid nozzle refilling.

Heat build up is another limiting factor in very high rate operations in viscous fluids and for certain ejector technologies. Highly viscous fluid requires more applied energy to eject with the upper limit set by fluid cavitation and physical damage to drop ejector structure. The energy applied to the working fluid in order to eject a microdrop using the same fluid can differ by orders of magnitude depending upon the drop ejector technology. Bubble jet (thermal inkjet) ejectors apply about a hundred times as much energy into the fluid per drop as reservoir wall displacement technologies such as piezoelectrically driven drop generators. The focused acoustic beam ejector technology, while having the advantage of being virtually immune to nozzle clogging, has the disadvantage of requiring about an order of magnitude more applied energy per drop than thermal inkjet devices.

In addition to a maximum drop ejection rate, some fluids have a minimum drop ejection rate. The physical mechanisms that can require that drops be ejected at a minimum rate are discussed in a previous section in reference to first drop reliability.

12.10 ENVIRONMENTAL AND TEMPORAL STABILITY

Some fluid mixtures do not have stable rheological properties as a function of time and environmental conditions. This can produce serious reliability problems. Some classes of fluids we have encountered that exhibited this type of behavior include:

- Unstable suspensions
- Unstable colloids
- Biological fluids containing live organisms
- Fluids containing volatile solvents
- Fluids that leave solid residues upon evaporation
- Non-pH buffered fluids containing calcium ions
- Highly viscous fluids

Hypothetical fluids that are likely to be problematic:

- Fluids operated close to their melting points
- Photoactive polymers

12.11 CHEMICAL REACTIVITY WITH DROP GENERATOR

If the fluid that one is using is corrosive to the materials making up the drop generator, then there will definitely be problems with long term stable operation. We worked with one research group that had this problem when they attempted to refill a commercial inkjet printhead with a biological payload suspended in a volatile solvent. The commercial printhead was not designed for use with organic solvents and came apart. Steven Zoltan's design of a tubular drop ejector, which is made of glass and has externally attached piezoelectric actuator, is a far better design for general purpose use in experimental science. The two solutions to this problem are to engineer a fluid that is nonreactive with the dropper and to utilize a dropper constructed from chemically nonreactive materials.

12.12 TOXICITY

Using the least toxic fluid components possible is an obviously desirable design criterion. Toxicity to both the experimenter and to biological payloads need to be considered. One additional hazard factor to take into account is that microdrop ejectors are capable of dispersing their operating fluid as an aerosol if overdriven.

12.13 COST

The cost of the test fluid or its limited availability can constrain the type of fluid drop ejector it can use for a particular application. Different fluid drop ejectors have widely varying minimum operational fluid volumes. Widely varying means from a few tens of milliliters to less than a microliter. Piezoelectric squeeze mode tubular reservoir ejectors have the smallest operational fluid volumes of the common commercially available microdrop generators.

REFERENCES

1. H.R. Kang, Water-based ink-jet ink. I. Formulation, *J. Imaging Sci.*, vol. 35, pp. 179–188, 1991.
2. H.R. Kang, Water-based ink-jet ink. II. Characterization, *J. Imaging Sci.*, vol. 35, pp. 189–194, 1991.
3. H.R. Kang, Water-based ink-jet ink. III. Performance Studies, *J. Imaging Sci.*, vol. 35, pp. 195–201, 1991.
4. R.E Drews, The effects of translationally symmetric nozzle face defects on the directional accuracy of thermal inkjet arrays, *IS&T's Seventh International Congress on Advances in Non-Impact Printing Technologies*, pp. 107–116, IS&T: The Society for Imaging Science and Technology, 1991.

CHAPTER 13

Components of Drop-on-Demand Ejector Fluids

In order to stabilize the fluid and obtain the rheologic properties needed for reliable operation, modern formulations for commercial inkjet fluids often contain a half dozen or more components.[1-4] Some of the additives serve more than one function. A commercial inkjet ink intended for document printing or deposition of reagents may contain fluid components consisting of:

- Payload
- Solvents
- Humectants
- Dispersants
- Surfactants
- Viscosity modifiers
- Polymeric fluid elasticity agents
- Anti-fungal agents (biocides, preservatives)
- Chelating agents
- pH controllers
- Corrosion inhibitors
- Defoamer
- Additives specific for thermal inkjet
  - Antikogation agent
  - Bubble nucleation promoter
- Additives specific for inkjet printing
  - Penetrants
  - Fixatives
  - Binder
  - Anticockel additive

- Ultraviolet blockers
- Free-radical inhibitors
- Antioxidants

Microdrop ejection fluids intended for use in research and manufacturing have different operational optimizations from commercial printing inks, but many of the basic rheological requirements, such as appropriate viscosity, surface tension, and particle content, remain constant for all microdrop ejection systems. Not all of the fluid components used in commercial printing inks are needed for nonimage printing applications, but the functions of each additive should be considered to verify its relevance or lack thereof to one's particular application.

13.1 PAYLOAD

This is the material that one wishes to put into a microdrop form in order to deposit, react, take a measurement with, or aerosolize. Some examples might be:

- Pigments and dyes for inkjet printing
- Solder for laying out solder bump pads
- Metals for conductive traces
- Nucleotides for DNA synthesis
- Cells and microorganisms
- Enzymes
- Proteins
- Antibodies
- Photoactivatable chemicals for medical therapeutics and dental work
- Volatile fluids for taking evaporation measurements
- Deposition of compounds to form organic LEDs
- Transparent optical material for microlenses
- Laser dyes
- Thermoplastics for droplet based manufacturing
- Deposition of ferrofluids for making micromachined magnetic sensors and actuators
- Solvents for precision microetching
- Covert labeling materials for document security
- Masking materials for micromachining processing
- Chemicals for combinatorial synthesis and testing
- Biological reagents for combinatorial testing
- DNA for making microarrays
- Fluids of different types for making rheological measurements
- Exotic solids such as moon dust for searching for subatomic particles with fractional charge
- Material to be vaporized by laser to form precise spectroscopy measurements
- Polymers for making microcapsules

Sometimes, as when ejecting liquid metals for soldering or oil for precision spot lubrication, the carrier fluid is the payload.

13.2 SOLVENT

This is the fluid that either suspends or dissolves the fluid drop payload. In some cases the fluid and the payload can be identical, such as liquid solder. Some common fluids that have been used as solvents in fluid drop ejectors include:

- Water
- Polyhydric alcohols such as:
  - Ethylene glycol
  - Propylene glycol
  - Trimethylene glycol
  - Glycerine
  - 1,3 Butanediol
  - 2,3 Butanediol
  - 1,4 Butanediol
  - Diethylene glycol
  - 1,5 Pentanediol
  - Hexylene glycol
  - Triethylene glycol
  - Dipropylene glycol
  - 1,2,6 Hexanetriol
  - 2-Propanol
- Alcohols
  - Ethanol
  - Propanol
  - Butanol
  - Cyclohexanol
- Ethers
  - Ethylene glycol monomethyl ether
  - Ethylene glycolmonoethyl ether
  - Triethyleneglycol monomethyl ether
  - Tripropyleneglycol monomethyl ether
  - Tripropyleneglycol monobutyl ether
- Alkanol amines
  - Monoethanol amine
  - Diethanol amine
  - Triethanol amine
  - N,N-dimethylethanol amine
  - Aminoethyl ethanol amine
  - Morpholine
- Dimethylsulfoxide
- Mercury
- Silicon oil
- Mineral oil
- Acetone
- Epoxies
- Methyl ethyl ketone
- Solder

The solvent typically makes up about 60 to 80% of the ejection fluid mix. Some solvents, such as the polyhydric alcohols, serve also as viscosity modifiers and humectants. It is common practice to mix different solvents together in order to obtain the desired rheological properties.

13.3 HUMECTANTS

Humectants are additives that are hygroscopic and miscible with the primary solvent. Aqueous based fluids with high vapor pressures can evaporate out when sitting idle in the ejection hole leaving either a solid crust clogging the hole or a viscous plug of fluid requiring a greatly increased drive amplitude to eject. The presence of humectants in the mixture prevents total drying out and solidification of the fluid in the ejection hole. Humectants are important in aqueous inks to prevent clogging of the ejection aperture and to minimize the severity of the first drop problem.[5] Some commonly used humectants are:

- Propylene glycol
- Glycerol
- Ethylene glycol
- Diethylene glycol
- Polyethylene glycol
- Polypropylene glycol
- N-methyl-pyrrolidine
- 2-Pyrrolidone
- N-methyl-2-pyrrolidone
- 1,2,-Dimethyl-2 imidazolidinone

Some of these humectants can also double as the principle solvent.

13.4 VISCOSITY MODIFIERS

There is an optimal viscosity range for fluid drop ejectors to produce reliable monodisperse drops at high ejection rates. Too high a viscosity and obviously the energy required for ejection becomes excessive, satellites due to ligament formation become difficult to avoid, and in some cases fluid cavitation effects are seen in the ejection fluid. Too low a viscosity and the fluid damping of meniscus oscillations may become insufficient, which limits the high frequency operation of the microdrop generator and in fact can make it very frequency sensitive. The viscosity range specified in patent literature as being optimal for reliable microdrop ejection is 1 to 30 cS.

One other reason for engineering the fluid to have a high viscosity is to extend the period of time that suspended solids will remain mixed and not settled out. In scientific research, it is sometimes not feasible to reduce the microdrop payload to the submicron dimensions required for the particles to remain stable in suspension.

In practice, viscosity is most often set by combining together compatibly miscible solvents with different viscosities, such as glycerol and water. Most of the viscosity modifiers in patent literature have been polyhydric alcohols, such as propylene glycol mixed in different percentages with water. In our lab, we have used light mineral oils mixed with heavier petroleum-based oils to produce nonaqueous fluids with the desired viscosities. Among the viscosity modifiers used are:

- Polyhydric alcohols in combination with water and other polyhydric alcohols such as
  - Ethylene glycol
  - Propylene glycol
  - Trimethylene glycol
  - Glycerine
  - 1,3 Butanediol
  - 2,3 Butanediol
  - 1,4 Butanediol
  - Diethylene glycol
  - 1,5 Pentanediol
  - Hexylene glycol
  - Triethylene glycol
  - Dipropylene glycol
  - 1,2,6 Hexanetriol
  - 2-Propanol
  - Alkyline oxide adducts of polyhydric alcohols

From Iwata et al. U.S. Patent 4,986,850:[6]

The polyhydric alcohols for the alkylene oxide adducts include tri- or higher-hydridic alcohols such as glycerin, hexose or its sugar-alcohol, pentose or its sugar-alcohol oligosaccharide etc. added with ethylene oxide, propylene oxide, butylene oxide, tretrahydrofuane or the like. The number of moles of alkylene oxide added per mole of a polyhydric alcohol is in the range from 1 to 50, preferably 2 to 25. The alkylene oxide added may be either a single species or plural species, and may be in the form of either a block copolymer or a random copolymer. The content of the alkylene oxide additive of the polyhydric alcohol is in the range of 0.05 to 4% by weight, and preferably 0.05 to 2% by weight base on the total weight of the recording liquid. The content less than 0.05% by weight will fail to exhibit the effect of the addition, while the content higher than 4% by weight will result in an excessive increase of the viscosity, causing a trouble of the discharge.

One caution about attempting to use static or low shear rate viscosity measurements to predict ejectability is that the shear rates in microdrop ejectors are in the 10^6 sec^{-1} range. This is orders of magnitude higher than the viscosity measurement range of rotational fluid viscosity measuring instruments. When ejecting complex fluids the dynamic viscosity can be very different from the static viscosity. Viscosity is defined as the resistance of a fluid under a force to flow. The viscosity of a fluid is not always independent upon the shear rate or forces applied to induce a flow.

There are four types of fluid behavior that characterize the relation between force, shear rate and effective viscosity.

13.4.1 Newtonian

Viscosity is independent of shear rate. Most carrier fluids for inkjet inks fall into this category of fluids. This means that in practice one can use the manufacturer's stated fluid viscosity curves to estimate ejectability.

Examples of Newtonian fluids would be:

- Water
- Motor oil
- Mineral oils
- Gasoline
- Mixtures of water and low molecular weight polyhydric alcohols
- In general, gases and low molecular weight fluids

13.4.2 Shear Thinning (Pseudo-Plastic)

Viscosity decreases with increasing shear rate. Most fluids that exhibit non-Newtonian behavior fall into this category. The general mechanism for the decrease in viscosity with increasing shear rate is the uncoiling and decoupling of entangled polymer chains, which under high shear can move independently and align with the direction of fluid flow. As a consequence of this basic mechanism, the reduction of viscosity with increasing shear rate is a characteristic of the fluid over a limited range of shear rates. At very high and very low rates, the viscosity tends toward asymptotes.

Examples of shear thinning fluids would be:

- Most paints
- Ballpoint pen ink
- Blood
- Some colloidal suspensions
- In general, fluid with long chain polymers

13.4.3 Shear Thickening (Dilatant Fluids)

Viscosity increases with increasing shear rate. This is a more uncommon behavior but is important for those researchers who may be attempting to eject fluid particle slurries. Attempts to eject this type of fluid can result in either no ejection at all or ejection of the carrier liquid minus the payload particles.

Examples of shear thickening fluid would be:

- Concentrated suspensions of cornstarch and water
- Wet sand
- Clay slurries
- Some colloidal suspensions

13.4.4 Plastic (Bingham-Plastic)

Flow does not occur until a force threshold is exceeded. Fluids exhibiting plastic flow behavior would likely be unejectable in drop-on-demand devices due to the lack of a capillary action induced flow allowing refill of the ejection region in the inkjet nozzle.

Examples of plastic fluids would be:

- Toothpaste
- Grease
- Tomato paste

13.5 SURFACTANTS

Surfactants are added to ejection fluids to alter the surface tension and fluid-to-dropper contact angles to more optimal values for stable droplet ejection. Optimizing the surface wetting characteristics of the fluid with different parts of the drop ejector is important. If the fluid does not wet the inside of the fluid reservoir and the ejection hole easily, drops may not form, air may be difficult to purge from the system during the fill operation, and fluid refill time of the ejection aperture may be slow enough to compromise frequency of operation.

On the other hand, the fluid should exhibit a high contact angle (difficult wetting) of the ejection aperture exterior surface. This is to suppress leakage of the fluid out through the ejection hole and the formation of a convex fluid meniscus that will block fluid jet ejection or render the ejection of drops unstable. Surfactants for this purpose are typically added in fractions of a percent of the total volume of the fluid. There are an extremely large number of manufactured surfactants that are selected because of their toxicity, reactivity with other fluid components, cost, stability, and chemical and physical range of conditions where their properties are specified to be stable. Some surfactants double as effective dispersants. Some surfactants listed in patent literature as being used in inkjet inks are:

- Nonyl-phenoxpolyethoxyethanol
- 3 M Corp. — Fluorochemical FC170C
- Union Carbide — Tergitol® 15-S-5
- Union Carbide — L-5340
- 3 M Corp. — FC-430

Typical surface tensions of inkjet inks are in the range 20 to 60 dynes/cm.

13.6 DISPERSANTS

Dispersants are additives that aid in the maintaining of solids such as pigment particles in stable suspension. They serve two practical functions. One is to aid in wetting. Structurally, the molecules facilitate this by having one portion that binds to the solid particle and another portion that is strongly philic with the fluid. The other function is to suppress agglomeration of the small particles into large, cohering masses, which will then fall out of suspension. They function by coating the particles with a surface layer that suppresses the van der Waal's mediated agglomeration of particles into larger solids, which would then gravitationally settle out or jam ejection aperture holes.

Dispersants have to be matched to both the solvent and the type of particles that one wishes to suspend. The two techniques used to stabilize suspensions are putting a charged surface layer on the particles (charge stabilization) or attaching log chain molecules to the surface of the particles (steric stabilization). Steric stabilization is less sensitive to failing due to interactions with other fluid components, is usable in both very high and very low particle loadings as well as being more usable in non-aqueous solutions. Sterically stabilized suspensions, however, have higher viscosities than an equivalent charge stabilized suspension. Some dispersants listed in the patent literature as being successfully used in inkjet fluids have included:

- Elementis Specialties, Inc. — Disperse-Ayd W-22 dispersant
- Union Carbide — Triton X-100® dispersant
- Rohm and Hass Company — Tamol® SN dispersant
- Kao Corp. — Emulgen® 420
- Avecia Limited — Solsperse® 27000
- Kao Corp. — Emulgen® A-90

13.7 POLYMERIC FLUID ELASTICITY AGENTS

Polymeric fluid elasticity agents are long, chain soluble molecules added to prevent satellite drop formation by suppressing fragmentation of the fluid jet into random sized spray.[7-9] These additives have noticeable effects on the drop formation process in concentrations as low as 10 ppm. The theory of operation is that the long chain molecules give the fluid a kind of elasticity which causes the fluid jet to have a greater tendency to remain in a cohering mass that ultimately pulls together into a single drop rather than disintegrating into many smaller separate satellite drops. Tests have shown long fluid jets that broke up into small drops without the additive, but with the additive formed continuous ligaments, which were either pulled into the primary drop or pulled the primary drop back into the orifice when ejection energy was insufficient to break the jet free. Some of the additives used for these tests were:

- Polyacrylamide (PAM,/-CH_2-C_2ONH_3-/$_n$) molecular mass — 500,000 to 6,000,000
- S.C. Johnson Polymer Ltd. — Joncryl® 680, 678, 67HPD 671, 586

13.8 ANTI-FUNGAL AGENTS (BIOCIDES, PRESERVATIVES)

The aqueous media used for many microdrop ejection fluids may deteriorate due to the growth of microorganisms. Most formulas for industrial inkjet inks have included some type of biocide to suppress the growth of bacteria, fungus and other microorganisms. The following biocides were listed as being used in various patented industrial ink formulations.

- Givaudan Corp. — Giv-Gard DXN anti bacterial agent
- Zeneca AG Products, Inc. — Proxel® CRL
- Huls America, Inc. — Nuosept® 95
- Olin Corp. — Omadines®
- Nopco Paper Technology Pty Ltd — Nopcocide®
- Bode Chemie — Bacillat® 35 preservative
- Nudex, Inc. — Nuosept®
- Union Carbide — Ucarcide®
- RT Vanderbilt Co. — Vancide®
- Sodium dehydroacetate
- Sodium benzoate
- Sodium pyridinethione-1-oxide
- Zinc pyridinethione-1-oxide
- 1,2-Benzisothiazolin-3-one
- Dow Chemical — Dowicil®4 75, 150, 200
- Sorbate salts

13.9 CHELATING AGENTS

Chelating agents are added to bind to metal ions to prevent the formation of scale deposits on evaporation of fluids near the ejection aperture. An important metal ion to suppress is calcium, which can combine with carbonate ions produced by absorption of atmospheric carbon dioxide to solid calcium carbonate deposits. Examples of chelating agents include:

- Ethylenediaminetetracetic acid (EDTA)
- Sodium salt of ethylenediaminetetracetic acid
- Diammonium salt of ethylenediaminetetracetic acid
- Tetrammonium salt of ethylenediaminetetracetic acid
- Sodium gluconate

13.10 pH CONTROLLERS

The pH of fluid jetting mixtures often needs to be controlled. One reason is that the surfactants and dispersants that prevent particle agglomeration may only function under certain pH values. The solubility of certain dyes may vary with pH. Fluids containing biological compounds obviously must be pH regulated to avoid killing live organisms or denaturing proteins. Some microdrop ejectors may be constructed from materials that may require a certain pH range to avoid corrosion. One subtle reliability issue requiring pH control of the fluid is if calcium ions are present in an aqueous fluid operated in air. If the pH of the fluid is uncontrolled, absorbed carbon dioxide will form carbonic acid and carbonate ions, which upon evaporation of the fluid near the ejection aperture may precipitate out solid calcium carbonate blocking off the ejection aperture hole with a hard scale deposit. Some of the pH buffers that were mentioned in inkjet formulation patents are:

- Sodium carbonate
- Aqueous solution of sodium hydroxide
- Aqueous solution of ammonia
- Lithium hydroxide
- Phosphate salts
- Diethanolamine
- Triethanolamine

13.11 CORROSION INHIBITORS

Many designs for microdrop ejectors have metal components that are subject to corrosion. Corrosion inhibitors added to the ejection fluid mixtures typically alter pH to inhibit reaction with the metals or deposit organic, thin film surface layers over the vulnerable surfaces.

13.12 DEFOAMER

Some designs for microdrop ejectors have active fluid circulation systems that can agitate the fluid in a way that stable masses of air bubbles are created and become intermixed in the fluid. This can cause serious reliability problems. Small air bubbles can absorb the pulse energy intended to propel the fluid. Another failure mode can result from the bubbles blocking off flow channels preventing fluid refill of the ejection chamber. The general mechanism that promotes foaming in fluid mixtures is the surfactant's interaction with thin, fluid films in a way that makes them resistant to rupture by stretching. Defoaming agents act to break up foams by destroying the integrity of their thin film walls by alteration of surface tension or the introduction of defects where film collapse is forced to occur. Some antifoaming agents mentioned in patents are:

- Kao Corp. — Antifoam® E-20
- Air Products and Chemicals, Inc. — Surfynol® 104

13.13 ELECTRICAL CONDUCTIVITY SALTS

This is primarily an issue with fluids that are used with continuous inkjet microdrop generators in which charge induction on the drops as they are produced is used to deflect them in flight with electric fields as a means of controlling material deposition. The materials mentioned in patent literature are typically nonprecipitating ionic salts such as potassium thiocynate.

13.14 ADDITIVES SPECIFIC FOR THERMAL INKJET

13.14.1   Antikogation Agent

Kogation is the formation of solid deposits over the heating elements in a thermal bubble actuated microdrop ejector. In order to vaporize the ink and produce the desired gas bubbles on the microsecond time frame required for fluid drop ejection, the local temperatures induced in the fluids range from 300°C to 500°C, and the power density at the heater has to be on the order of 10^9 watts/meter.[2] As a result of these extreme local conditions, there is decomposition of the components of the fluid that often results in solid deposits forming over the heating elements. These solid deposits act as thermal insulators, which reduce the efficiency of the ejector over time, and, if extreme, halt drop production. Antikogation formulations have utilized chemistries that replaced sodium salts, which research has shown to promote kogation with lithium, ammonium, or alkyl ammonium cations. Certain humectants, such as 2 methyl 1,3 propanediol, were claimed to suppress kogation. Oxo anions such as phosphates, polyphosphates, and phosphate esters have been added to reduce the incidence of kogation. Other compounds claimed to reduce kogation include organic acid sulfonates, such as sodium methane sulfonate, sodium, 4-toluene-sulfonate and sodium propene-1-sulfonate.

13.14.2   Bubble Nucleation Promoter

As the temperature of a fluid rises, the temperature at which vaporization starts can shift. To stabilize this process, compounds such as polyethylene oxide, alkoxy-polyalkyleneoxyalkanol, and polyorganosiloxane have been added to inkjet ink formulations. Other strategies have been to utilize an emulsion consisting of a small percentage of volatile fluid with a lower vapor pressure than the primary solvent. Another approach utilized modified pigment particles in a specified size range as bubble nucleation points.

13.15 ADDITIVES SPECIFIC FOR INKJET IMAGE PRINTING

Penetrants are additives that aid in the penetration of the ink into fibrous media such as paper or fabrics. Without this type of additive, some inks may bead up on the surface of the paper, suffer smearing, and have an excessively long drying time.

Mechanical immobilization of the dye of pigment, once it is deposited on the image, is important for documents that may be frequently handled or stacked and rubbed against other documents. Fixatives and binders are additives, typically resins and polymers, which are intended to increase the smear resistance of the printed image.

Anticockel additives reduce the tendency for the ink, when absorbed into the paper, to wrinkle, curl, or otherwise mechanically distort the final document.

Additionally, ultraviolet blockers, free-radical inhibitors and antioxidants are utilized to protect the image from fading. A fundamental problem with all forms of printing is image permanence. This is commonly seen in more prosaic conditions by sunlight causing paints to fade. The three main causes of chemical degradation of dyes and pigments are exposure to light, particularly the ultraviolet components of sunlight; atmospheric oxygen; and chemical free radicals. There is a large amount of research presently being conducted to find methods of enhancing color permanence of inkjet printable inks.

REFERENCES

1. H.R. Kang, Water-based ink-jet ink. I. Formulation, *J. Imaging Sci.*, vol. 35, pp. 179–188, 1991a.
2. H.R. Kang, Water-based ink-jet ink. II. Characterization, *J. Imaging Sci.*, vol. 35, pp. 189–194, 1991b.
3. H.R. Kang, Water-based ink-jet ink. III. Performance Studies, *J. Imaging Sci.*, vol. 35, pp. 195–201, 1991c.
4. S.F. Pond et al., Ink Design, in *Inkjet Technology and Product Development Strategies*, Stephen F. Pond, Ed., Torrey Pines Research: Carlsbad, CA, 2000, Chapter 5.
5. J. Dressler, Two-dimensional, High Flow, Precisely Controlled Monodisperse Drop Source, Gov. Doc. WL-TR-93–2049, Aero Propulsion and Power Directorate, Wright Laboratory, Air Force Material Command, 1993.
6. K. Iwata, S. Tochihara, and S. Koike, Recording liquid:water composition with 25 cp minimum viscosity at 9:1 blend and 15 cp maximum viscosity at 1:1 blend, U.S. Patent 4,986,850, 1991.
7. A.V. Bazilevsky, J.D. Meyer, and A.N. Rozhkov, Effects of polymeric additives on vapor bubble dynamics in thermal inkjet printing, *Recent Progress in Ink Jet Technologies II*, Society for Imaging Science and Technology, Springfield, VA, pp. 291–294, 1999.
8. C.M. Evans et al., Optimization of ink jet droplet formation through polymer selection, IS&T's NIP 15: International Conference on Digital Printing Technologies, Orlando, FL, pp. 78–81, 1999.
9. J.D. Meyer, A.V. Bazilevsky, and A.N. Rozhkov, Effects of polymeric additives on thermal ink jets, Recent Progress in Ink Jet Technologies II, Society for Imaging Science and Technology, Springfield, VA, pp. 450–455, 1999.

CHAPTER 14

Making Jettable Suspensions of Ground Solids

Some proposed payloads for microdrops include ground up solids to be suspended in a carrier fluid and then jetted out of the drop ejector. Despite its successful use in inkjet printing, this is technically a very difficult engineering task if one's goal is to take an arbitrary solid, fragment it, suspend it in a jettable fluid, and eject it reliably on demand from a microdrop generator.

The first problem is that the maximum diameter for a particle that can be stably suspended depends upon the material, and the fluid must be between 0.1 and 1.0 micron in diameter. Larger particles will gravitationally settle in a timeframe too short for most applications. In addition, as the size of the particles approaches the diameter of the ejection aperture hole, the presence in random locations of large particles unpredictably destabilizes the ejected fluid jet leading to unreliable drop formation.

Conventional mechanical crushing and grinding becomes ineffective for reducing particle size when the particle diameters are in the 10- to 100-micron diameter range. This is primarily for two reasons. The first is that solids subjected to mechanical stress fracture preferentially along lines of weakness, such as crystal lattice defects. As particles are fractured into smaller sizes, the number of such defects drops. The force required to fracture a defect-free small particle is far higher than that needed to break it along a crystal plane dislocation. The second problem is reagglomeration. Small, freshly fractured particles with clean crystal planes that contact each other in the right manner can fuse back together into a single particle.

Once dispersed in a fluid, van der Waals forces can still cause even otherwise stably suspended particles to bind together with different degrees of adhesion. A large aggregate particle can gravitationally fall out of suspension or jam an ejection hole. There is another failure mode involving the formation of loosely bound together masses of suspended particles known as particle flocculation. As an example of loose aggregation, the Microdrop Particle Search Group once made a mixture of 50-

nanometer diameter aluminum oxide powder in silicone oil, which looked like it was stably suspended. Upon attempting to eject it from a microdropper, nothing but clear fluid was ejected. Apparently, the aluminum oxide powder had flocculated together in a loose aggregation of the fine particles. These particle aggregates acted like a loose porous sponge-like mass that allowed the carrier fluid to flow through and be ejected but kept the solid suspended powder in the drop ejector.

Despite the practical difficulties, it is possible to make stable suspensions of ground solids capable of being jetted as evidenced by the successful development of pigmented inkjet inks. The steps needed to take a solid and place it in suspension are:

- Obtaining small diameter particles of the desired solid payload
- Wetting and dispersing of these solid particles in the carrier fluid
- Filtering or settling to remove oversized particles capable of clogging fluid ejection nozzles

14.1 OBTAINING SMALL DIAMETER PARTICLES

In order to make a stable jettable suspension, the solids must be less than a micron in diameter, preferably closer to 0.1 microns in diameter for high specific gravity solids. However, it is not necessary that all of the solids be in this diameter range. The larger diameter particles can be taken out of the suspension later by filtering or through settling. The payload of the microdrops will be composed however only of the smaller size particles.

There are two ways of obtaining small particles of a given material. One is to synthesize them from the dissolved ions or vapor phase compounds and precipitate out the solids from vapor or liquid phase in the size range that one desires. This is the way that most pigments are made. For scientific applications the more common case will be that there is a specific solid mineral that one will want to place into suspension for analysis. Synthesizing it from atomically dispersed raw elements is not an option. In addition the quantities of the solid that one starts with may be limited so that one cannot use industrial methods that will result in a loss of a large amount of the base material.

We had the problem of suspending carbonaceous chondrite meteorites and fluoraptite in jettable fluid mixtures. We solved the fragmenting problem in the following manner.

First the mineral is fragmented by conventional grinding, which for small samples can be crushing with a hammer and anvil followed by grinding into a fine powder with a mortar and pestle (Figure 14.1). At the conclusion of this stage, the powder is reduced to particles primarily between 10 to 100 microns in diameter.

Air jet milling is used to reduce this relatively coarse powder to submicron particle diameters. An air jet mill functions by taking coarsely ground particles in the 1- to 100-micron range and accelerating them to supersonic speeds with high pressure air. The particles are accelerated through internal nozzles to supersonic speeds and collided with each other. The resultant submicron fragments are collected in a container under the milling chamber. The device shown here is a commercial air jet milling apparatus (Figure 14.2). This particular model is espe-

Figure 14.1 Crushing a fragmentable solid using a mortar and pestle. The initial fragmenting of materials solids for the purpose of producing an ejectable colloidal suspension can be implemented by a process as simple as mechanical crushing with a mortar and pestle. The typical mean particle diameters achievable by mechanical crushing is in the 10–100 micron range.

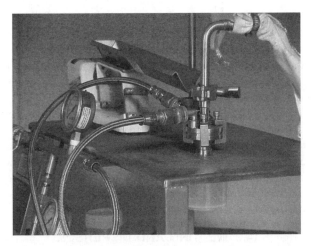

Figure 14.2 Air jet milling apparatus. Pressurized air is used to accelerate mechanically crushed solids to supersonic speeds and collide them with each other internally to produce powdered solids in the sub-1-micron diameter range.

cially suitable for scientific experimental use due to its small size, which is necessary to minimize the loss of pulverized solids by adhesion to processing chamber interior side walls.

14.2 NECESSARY PARTICLE SIZE

The diameter that it is necessary to reduce the particles down to in order to obtain a stable colloidal suspension varies with the specific gravity of the fluid and solids as well as the viscosity of the fluid. It is possible to obtain a rough estimate of the size reduction required by calculating the effective Stokes law limited settling rate and comparing this value to the displacement per unit time by Brownian Motion.

$$k = \text{Boltzmann's constant } (1.38 \times 10^{-16} \text{ erg/Kelvin}) \tag{14.1}$$

where η = fluid viscosity
     d = drop diameter
     σ_f = fluid density
     σ_p = particle density
     T = temperature
     g = gravitational acceleration
     t = time

Stokes Law limited settling speed:

$$v_{\text{Stokes}} = [d^2 (\sigma_p - \sigma_f) g]/18\eta \tag{14.2}$$

Particle displacement per unit time due to Brownian Motion:

$$\Delta x_{\text{Brownian}} = [(2 k T t)/(3 \pi \eta d)]^{1/2} \tag{14.3}$$

As a rule of thumb, if the particle displacement by Brownian motion is greater than that predicted from the Stokes Law settling rate then it can form a stable colloidal suspension, otherwise the mixed particles will rapidly settle out. For particles with the density of most minerals (~20 grams/cm^3) suspended in water the maximum particle diameter is about 1 micron.[1-4] Commercial inkjet inks have pigment particle diameters that range from 0.1 to 0.2 microns. Note that while the ratio of the predicted displacement by Brownian motion relative to the gravitational settling rate can be used as a crude measure to predict stability, environmentally induced convective mixing is the primary reason for the commonly observed long term stability of most colloidal suspensions.[3,4]

14.3 WETTING AND DISPERSING

The powder produced by the grinding and air jet milling operations must next be placed into suspension in the carrier fluid. The particles after being fragmented and likely to be adhering together in agglomerated masses that require dispersion before the individual particles can be thoroughly wetted by the fluid. The fluid must in turn contain surfactants to coat the surfaces of the particles to suppress reagglomeration and settling, and depending upon the contact angle of the carrier fluid with the particles a surfactant wetting agent may have to added to aid in having the particles completely wetted by the fluids. To break up the particle agglomerations and wet the surfaces of the fine powder requires that the particle agglomerates be subjected to a high shear force while in the presence of the surfactant loaded carrier fluid. There are many methods of accomplishing this dispersing task. For small fluid volume work in research environments, the techniques most readily applicable are:

14.3.1 High Powered Ultrasound

Both ultrasonic probes and vessels externally excited by ultrasound can be used for dispersing powdered solids if the ultrasonic intensity is sufficient to produce sufficiently high levels of convection and cavitation in the fluid without overheating the fluid or breaking down the constituent additives.

14.3.2 Fluid Jet Impingement

There are commercial devices that take particle-containing fluids and collide high-speed jets of these fluids together to break up particle agglomerations. A variation on this technique recirculates fluid through a region with a flow constriction generating very high changes in local fluid velocity to produce a high local shear rate to break up agglomerated particle masses.

14.3.3 Rotary Mixers

The lowest cost dispersing device that operates easily with the small volumes of fluids are the rotary homogenizers (Figure 14.3).

The rotary homogenizer generates high shear mixing by using high speed rotating blades moving fluid in close proximity to the inner walls of a cylinder containing

Figure 14.3 Rotary homogenizer (left) and details of the high shear rate mixing tip (right). High speed rotation of the rotor inside the slotted stator draws the fluid powder mixture into the inside of the stator where it is exposed to a high shear region between the rotor and the stator and is forced out the slots. The fluid shear between the rotor and stator accomplishes the necessary particle deagglomeration, wetting and dispersing.

narrow fluid passages. The fluid is drawn into the center of the rotor and then subjected to high sheer forces in the narrow gap between the rotary mixing blades and the inner walls of the containment cylinder. The fluid pressure gradients are high enough that we have observed fluid cavitation taking place at high RPMs. We have found that the period of mixing required for effective dispersion and wetting is on the order of one to several hours for 10- to 20-ml volumes of fluids.

There was a very striking difference in the quality of the suspension and the method used for mixing and dispersion. As an example, a fluid suspension of carbonaceous chondrite meteorite suspended in light mineral oil using 2% Castrol motor oil as a source of surfactants and dispersants was mixed by hand by simply vigorously shaking a test tube containing these fluid components by hand. This fluid mixture settled out over the course of a few days leaving clear fluid on top of a bottom layer of a thick mud like black mass. In contrast, the same formulation mixed for 12 hours in a rotary homogenizer has remained opaque for months after the initial mixing. We have found that the method and duration of mixing make a strong difference in the quality of the final suspension.

The amount of motor oil added to the mineral oil as a source of surfactants was also critical to success of the suspension as a drop-on-demand fluid. A zero percent mix of motor oil resulted in a settling out of the powdered solids immediately after mixing ceased. A one percent mix lasted overnight until settling occurred and left clear fluid over a sharply defined layer of settled out solids. A two percent addition of motor oil resulted in a stable suspension that was readily ejectable. Five and ten percent by volume additions of motor oil resulted in stable suspensions that were not ejectable as monodisperse drops due to the formation of long connecting ligaments from the drops to the ejection nozzle.

14.4 FILTERING AND SETTLING

Particle suspension made from materials manufactured by synthesizing the solids from vapor of liquid phase components can often have very narrow particle size ranges. In contrast, a dispersed powder made from breaking up a large, nonhomogenous solid will likely result in a wide dispersion of particle sizes depending upon the friability of each component and each local region of the materials making up the solid. The practical implication for this is that the suspension prior to being used in a fluid drop ejector must have some way of keeping the large particles from clogging the ejection hole or disrupting the ejection process. One method of excluding large particles is to allow the suspension to settle out the larger particles and then draw off only the top portion of the fluid mixture for use in the fluid ejection hardware. Some designs for fluid drop ejectors have included particulate screens in or near the ejection region to keep large particles away from the ejection aperture. Filtering is possible if the filter element size can be sufficiently larger than the mean particle diameter that the filter does not clog and start filtering out the smaller diameter particles that one wishes to retain in suspension.

When making suspensions of solids, the maximum mass fraction that is possible to suspend and eject must be determined. These are two separate issues. Through experiments, members of the SLAC research group found that for each material, there is a limit to the maximum mass fraction of solids that it is possible to place in the liquid before the suspension collapses. For instance, in attempting to make suspensions of ground carbonaceous chondrite meteorites in light mineral oil, a 20% by weight suspension was apparently indefinitely stable but a 40% by weight mixture fell out of suspension in minutes, leaving clear fluid following termination of mechanical dispersing.

Apparently beyond a certain particle density in a given volume of liquid, the particles bind together and are dragged down as a unit taking with them submicron particles that would otherwise be suspendable indefinitely. On the other hand, a 40% by weight mixture of fluoraptite mineral in ethylene glycol was stable over a period of weeks. Each combination of fluids and suspended particles must be evaluated experimentally. As a rough guide to the proportions that are likely to be successful, commercial pigment based inkjet inks typically have a 5 to 10% by mass fraction payload of suspended pigment particles.[5]

14.5 ESSENTIAL REFERENCES

There is an enormous amount of literature on the subject of suspending solid particles in fluids of which this short summary cannot pretend to do justice. What is recommended in addition to perusing this literature is to contact representatives from the major chemical manufacturers to obtain dispersants, surfactants, and application notes for specific solids and carrier fluids. A patent database search will yield hundreds of a patents with relevant technologies.

There is one essential reference to obtain for researchers who need to place ground solids in suspension: Preparation of suspensions for particle size analysis, methodical recommendations, liquids and dispersing agents by Claus Bernhardt, *Advances in Colloid and Interface Science*, 29, pp. 79-139, 1988 Elsevier Science Publishers B.V., Amsterdam, Netherlands.[6]

This 60-page journal article details the theory behind stable suspensions and contains a 24-page long table of materials ranging from natural minerals, manufactured chemical compounds, to solids of biological origin along with recommended additives to assist suspension in common fluids.

14.6 EJECTABILITY

The ejectability of a suspension is difficult to impossible to predict from the fluid components alone. As an example, an experiment at SLAC to search for particles with fractional electric charge in meteoric materials is using a mixture of light mineral oil and motor oil and fragmented carbonaceous chondrite meteorites as a fluid to eject as microdrops. The mineral oil alone ejected acceptably. The mineral oil mixed with motor oil produced a chaotic spray of randomly sized drops. The total mixture of mineral oil, motor oil, and 6% by weight meteorite powder ejected more reliably than the pure mineral oil. This result is not something that was predictable without actual testing of the final fluid mixtures. It is important to test the final fluid mixture and not try to predict performance from the ejectability of the individual fluid components when tested alone.

14.7 FINAL CAUTION — MAKE SURE YOU HAVE A REAL SUSPENSION

One final caution is that there is a big difference in the behavior of a stable colloidal suspension and a fluid particle slurry. A fluid particle slurry is akin to mud or wet sand. In a colloidal suspension the suspended particles are both kept from agglomerating with each other and are bound to the fluid by surfactant coatings. For instance, a colloidal suspension in a test tube that is tipped and righted will have the portion of the glass that wetted the sides be clean of debris once the fluid has settled back down. A slurry will leave particle debris behind. A slurry can appear to be stably suspended where in fact the particles have formed a loose porous agglomerate or floc. If this kind of fluid does not clog the ejection hole under ejection testing, it will jet out the carrier fluid leaving the porous mass of loosely adhering particles behind. This is analogous to the process where mud and wet sand obtain their shear thickening non-Newtonian behavior. When attempting to move rapidly through wet sand, the fluid is compressed out of the pores of the sand bringing together sand grains whose mutual friction impedes movement through the now effectively drier wet sand slurry.

REFERENCES

1. E. Kissa, *Dispersions — Characterization, Testing, and Measurement,* Sedimentation, chap. 9, Marcel Dekker, New York, pp. 345–398, 1999.
2. Th.F. Tadros, Settling of suspensions and prevention of formation of dilatant sediments, in *Solid/Liquid Dispersions,* Th.F. Tadros, Ed., Academic Press, London, 1987, chap. 11.
3. H.N. Stein, *The Preparation of Dispersions in Liquids,* Marcel Dekker, New York, pp. 29–31, 1996.
4. I.D. Morrison and S. Ross, *Colloidal Dispersions — Suspensions, Emulsions, and Foams,* Kinetic and Statistical Properties, chap. 3, John Wiley & Sons, New York, pp. 48–61, 2002.
5. K. Suzuki, K.Koseki, and T. Amari, Dynamics of droplet formation in ink jet printer, *Recent Progress in Ink Jet Technologies II,* Society for Imaging Science and Technology, Springfield, VA, pp. 295–298, 1999.
6. C. Bernhardt, Preparation of suspensions for particle size analysis, methodical recommendations, liquids and dispersing agents, *Adv Colloid & Interface Sci.,* vol. 29, pp. 79-139, 1988.

Setting Up a Microdrop System ASAP

One common question often asked is if one is given the task of assembling a microdrop generation system as soon as is possible how should one go about it and what is the approximate hardware cost?

The components needed whether purchased or constructed in-house are:

Dispenser — the generic name for the nozzle and actuator assembly
Driver — the electronics needed to run the dispenser
Imaging — hardware and electronics needed to image the microdrops
Misc. — fluid pressure control, fill and purge pumps, mechanical mounting hardware

One item that is essential for drop dispenser maintenance is a bidirectional variable speed peristaltic fluid pump such as those manufactured by Masterflex®. Set up with a venting valve and a combination vacuum-pressure gauge, this pump is used to fill drop ejection systems, facilitate cleaning operations and assist in the unclogging of jammed nozzles.

I.1 BUYING A SYSTEM

The easiest way to acquire hardware if one has the money and one's task is within the limits of what is commercially available is to purchase a ready to use system.

The order of magnitude cost for various systems components are:

Dispensers — approximately $500/each.
Drive electronics — for the dispensers can be purchased commercially for $3,000–$5,000.
Imaging systems — costs are highly variable depending upon the intended application. A computer interfaced, strobed illumination machine vision system suitable for microdrop imaging can be purchased for about $10,000. A fiber optic coupled halogen illuminator suitable for backlighting microdrops to allow naked eye viewing costs about $300.

I.2 COMPANIES THAT SELL NONIMAGE PRINTING MICRODROP GENERATING SYSTEMS

I.2.1   Piezoelectrically Actuated Drop-on-Demand Systems

MicroFab Technologies
1104 Summit Avenue
Suite 110
Plano, TX 75074
www.microfab.com
972–578–8076

GeSiM Gesellschaft für Silizium-Mikrosysteme mbH
Bautzener Landstrasse 45
D — 01454 Grosserkmannsdorf
Germany
www.gesim.de
+49-(0)351–2695–322

Engineering Arts
7236 NE 91st Avenue SE
Mercer Island, WA 98040
www.engineering-arts.com
206–275–3855

Picojet
3155 SW 234th Avenue, Bldg. C
Hillsboro, OR 97123
www.picojet.com
503–356–0598

Perkin-Elmer
Life Science Headquarters (PiezoTipnology)
549 Albany Street
Boston, Massachusetts 02118
lifesciences.perkinelmer.com
617–482–9595

I.2.2   Focused Acoustic Beam Drop-on-Demand Systems

Picoliter, Inc.
231 South Whisman Road
Suite A
Mountain View, CA 94041
www.picoliterinc.com
650–940–4570

I.2.3   Fluid Displacement Drop-on-Demand Systems (TopSpot® Array Printer)

HSG-IMIT
Wilhelm-Schickard Srt 10
D-78052 Villing-Schwenningen
Germany
www.microfuidics.de
+49–77 21 9 43 271

I.2.4   Continuous Jet Microdrop Arrays Printers

Videojet Technologies, Inc.
1500 Mittel Boulevard
Wood Dale, IL 60191
www.videojet.com
630–860–7300

I.2.5   Imaging Systems

Imaging Technology International Corp.
4858 Sterling Drive
Boulder, CO 80301
www.inkjetsys.com
303–443–1036

VisionJet
Lumen House
Lumen Road
Royston
Herts SG8 7AG UK
www.visionjet.com
+44 1763 249444

Videojet Technologies, Inc.
1500 Mittel Boulevard
Wood Dale, IL 60191
www.videojet.com
630–860–7300

I.2.6   Microdrop Systems Consulting

MicroFab Technologies
1104 Summit Avenue
Suite 110
Plano, TX 75074
www.microfab.com
972–578–8076

Xennia Technology, Ltd.
Lumen House
Lumen Road
Royston
Herts SG8 7AG. UK
www.xennia.com
+44 (0)1763 246600

I.3 BUILDING A SYSTEM

One's situation may require the in-house construction of a microdrop system. One's requirements may be unique enough that no off-the-shelf system is suitable, or perhaps for cost and delivery time issues it is necessary to be able to fabricate and customize critical components such as droplet dispensers in-house. At $500 each, microdrop dispensers can become a very large budget item if one needs a large range of nozzle diameters, is constructing large element arrays, or if one's fluids are prone to producing unclearable clogged nozzles.

I.3.1   Drop Dispensers

The easiest to use and lowest cost to fabricate microdrop ejectors that we have worked with is a Pasteur pipette with its tip thermally formed into an ejection nozzle. An epoxy attached piezoelectric disc provides the actuation.

The required hardware for heat forming ejection nozzles includes:

Binocular inspection microscope	$500
Rotary hand grinder such as a Dremel Mototool	$100
Battery powered hand drill	$100
Propane torch	$30
Microscope with calibrated reticle	$500
Cytonix fluorocarbon fluid PFC 1601V	$140

Researchers at the Microdrop Particle Search team purchased piezoceramic actuators from:

APC International Ltd.
Duck Run
P.O. Box 180
Mackeyville, PA 17750
www.americanpiezo.com/
570–726–6961

The part number for the disc shaped drivers is R-1.00-.250-.100–855, gold plated contacts. The piezoelectric discs range from $10 to $20 each depending upon quantity ordered. The lead time is 6 to 8 weeks.

The part number for the cylindrical drive elements is C-.200-.100-.630–855. These piezoelectric cylinders cost $25 to $35 dollars each depending upon quantity ordered.

The fluoropolymer fluid FluoroPel® PFC 1601V for producing a high contact angle exterior nozzle surface is manufactured by:

Cytonix Corporation
8000 Virginia Manor Road
Beltsville, MD 20705
www.cytonix.com
301–470–6267

The cost of the FluoroPel® PFC 1601V fluid is $140 per pound.

The directions for forming the tips into ejection nozzles are detailed in Chapter 9. The easiest diameter nozzles to work with range from 25 to 75 microns in diameter. Smaller nozzles are difficult to clean, hard to fill and are prone to clogging during use. Larger diameter nozzles have difficult meniscus control problems. In other words they tend to leak fluid out the tip and are prone to ingesting air. Our preferred diameter nozzle for experimental work in which the microdrop diameter is noncritical is 35 microns. A glass nozzle coated externally with the Cytonix PFC 1601V fluid will function with aqueous fluid without requiring a pressure control manometer.

I.4 DRIVE ELECTRONICS

The minimum drive bandwidth requirements is 1 MHz in order to have adequate drive pulse rise and fall times. The amplitude required is in the low tens of volt range for low viscosity fluids such as water up to the hundreds of volt range for viscous suspensions. A commercial RF linear power amplifier driven by a triggerable pulse generator can drive the drop dispenser. Though a crystal controlled digital pulse generator is preferred analog pulse generators will work adequately to set the drive pulse widths. Reconditioned digitally programmable pulse generators sell for $1500 and up. Analog pulse generators sell for about $500. The cost of a new linear power amplifier suitable for driving drop ejectors is in the $2000 range. The schematics for a number of low cost circuits able to drive piezoelectrically actuated drop dispensers are presented in Chapter 8.

I.5 IMAGING

The recommended imaging system for an experimenter who has never before operated a microdrop ejection system is a fiber optically couple halogen lamp illuminator used to backlight the microdrops for direct visual observation. Stroboscopic machine vision systems if synchronized poorly to the drop ejection electronics can easily fail to register ejected microdrops. Many of our laboratory systems have both a fiber optic illuminator for getting a rapid naked eye evaluation of the behavior of a drop ejector as well as a computer interfaced machine vision system.

I.6 SETUP AND TESTING

I.6.1   Cleaning

The drop ejector fluid reservoir and fluid must be free of particles large enough to block the ejection aperture. After fabrication the inside of the drop dispenser is typically filled with dust and grinding debris. This debris can easily be seen by back illuminating the drop ejector and inspecting it with a binocular inspection micro-scope. To clean the dispenser, first draw particle filtered distilled water through the tip of the drop ejector utilizing a peristaltic pump to draw the vacuum. The vacuum tubing is then removed from the rear of the drop ejector and the drop ejector is pressurized by placing over the tip a small diameter tube containing a source of pressurized particle filtered gas such as canned inert dusting gas used to blow off optics. This is used to blow the cleaning liquid out the back of the drop ejector.

The process is repeated until visual inspection of the drop ejector shows no large particles. The back end of the drop ejector can be particle sealed with a rolled up piece of lint free optical cleaning paper. One test for the efficacy of the cleaning process is to draw from the tip under vacuum in a quantity of liquid equal to the amount that will fill the drop ejector during actual use and then to reverse the peristaltic pump and drive the fluid out of the tip as a continuous fluid jet. If the drop ejector can be repeatably filled and emptied in this manner without stoppages then the interior can be considered clean enough for reliable operation. The residual water remaining can be evaporated out by applying a vacuum to the drop ejector using the peristaltic pump.

I.6.2   Filling

The preferred fluid for initial system testing is a particle filtered mixture of 75% propylene glycol and 25% distilled water. This fluid mixture is nontoxic and ejects far more reliably that pure water. The fluid should be drawn in from the tip by placing the drop dispenser under a vacuum. The fill level should be about 1 to 2 cm past the mounting point of the piezoelectric drive disc. If the tip of the drop ejector has been externally fluorocarbon coated using the Cytonix PFC 1601V fluid then no external pressure control tubing will be needed for low ejection rate operation.

I.6.3   Electronic Drive Settings

Drive the piezoelectric element using a single monopolar pulse of about 5 microseconds width. The repetition rate should be about 100 Hz. This ejection rate is fast enough to produce an easy to observe drop chain, high enough in frequency to be audible to facilitate a check on whether the drive pulses are reaching the piezoelectric element and is slow enough that surface wetting effects and undamped structural resonances are not operational problems. Backlight the tip of the drop ejector using the fiberoptic illuminator. Increase the drive amplitude until microdrops appear. Adjust amplitude and pulse width until monodisperse drops of the desired diameter are produced.

Example Inkjet Ink Formulations
from Patent Literature

There are hundreds of inkjet ink formulations that have been the subject of patents. The reasons mentioned in patents for the selection of ingredients and the determination of their proportions were useful in determining the general make up of fluids suitable for microdrop ejection but one gets the overall impression given the large number of diverse formulations patented by different companies to perform the same task in which:

- There is a lot of trial and error involved (i.e., no exact science).
- There are a lot of different mixtures that work.

The inkjet formulas given in patent literature are useful as starting points in selecting fluid components and in determining the approximate proportions to use based on what researchers have found worked for their applications in the past. One factor to be noted is that the mixing process, particularly the conditions under which solid pigments were dispersed can be critical to the stability of the final mixture. These details are often not mentioned in the patent disclosures. One other caution to be noted in that while the inventors are required to include enabling information in sufficient detail to allow another worker skilled in the art to replicate the inventors device, patent law does not prevent the inventors, or more accurately the inventors' company intellectual property lawyers, from including in the patent disclosure large volumes of suboptimal alternative approaches that one may cynically believe are meant to slow down competitors by providing false leads.

II.1 DYE BASED INKJET INKS

U.S. Patent Number 5,076,843
Nonaqueous Thermal Inkjet Compositions
Acitelli et al.
Lexmark International, Inc.
December 31, 1991

Intraplast® CN dye	5%
Ethylene glycol	55%
2-Propanol	40%
Intraplast® CN dye	5%
Ethylene glycol	45%
N-methyl-2pyrrolidone	10%
Ethanol	40%
Nepton Black FF dye	2.5%
Ethylene glycol	57.5%
2-Propanol	40%
Intraplast® CN dye	5%
N-Methyl-2pyrrolidone	75%
2-Propanol	20%
Intraplast® CN dye	5%
Ethylene glycol	55%
Ethanol	40%
Iosol® Black dye	5%
Diethylene glycol	75%
Methanol	20%
Iosol® Black dye	3%
Ethylene glycol	57%
2-Propanol	40%
Solvent Black 7 dye	3%
Ethylene glycol	57%
2-Propanol	40%
Nepton Black FF dye	3%
1,4 Butandiol	57%
Methanol	40%
Intraplast® CN dye	5%
Carbowax 200	20%
2-Propanol	75%

U.S. Patent Number 4,239,543
Noncrusting Jet Ink and Method of Making Same
Beasley
Gould, Inc.
December 16, 1980

DNX = 10% by weight in water Giv-Gard DXN antibacterial agent

Ethylene glycol	85%
water	12%
dye	3%

Ethylene glycol	83.3%
deionized water	13.2%
Cyanimid Calco-Nigrosine WSB dye	2%
DXN antibacterial agent	1.5%

Propylene glycol	53.0%
deionized water	45.0%
Cyanimid Calco-Nigrosine WSB dye	2.0%

hydrochloric acid to adjust pH and fungicide as needed

U.S. Patent 4,853,037
Low Glycol Inks for Plain Paper Printing
Johnson et al.
Hewlett-Packard Company
August 1, 1989

dye	1–4%
Ethylene glycol or Diethylene glycol	5–10%
biocide (Proxel® CRL, Nuosept® 95)	0.01–0.3%
pH buffer	0.05–0.1%
(Sodium borate, Sodium hydrogen phosphate, Sodium dihydrogen phosphate) deionized water	balance of formula

U.S. Patent Number 4,285,727
Ink Compositions for Ink Jet Recording
Uehara and Itano
Konishiroku Photo Industry Co., Ltd.
August 25, 1981

water soluble acid or direct dye	0.5–8%
polyhydric alcohol such as	45–70%

- Ethylene glycol
- Propylene glycol
- Trimethylene glycol
- Glycerine

- 1,3 Butanediol
- 2,3 Butanediol
- 1,4 Butanediol
- Diethylene glycol
- 1,5 Pentanediol
- Hexylene glycol
- Triethylene glycol
- Dipropylene glycol
- 1,2,6 Hexanetriol

pH buffer   0.1–5%
- Potassium carbonate

preservative (biocide)   0.01–0.5%
- Bacillat® 35

surface active agent such as   0.05–0.5%
- Union Carbide L-5340
- 3M FC-430

chelating agent such as   0.1–1%
- Sodium gluconate
- Ethylenediaminetraacetic acid
- Disodium ethylenediaminetetraacetate
- Trisodium ethylenediaminetetraacetate
- Tetrasodium ethylenediaminetetraacetate
- Sodium diethylenetriaminepentaacetate

Ethylene glycol	43.7%
Diethylene glycol	10.0%
dye	2.5%
distilled water	33,5%
Potassium carbonate	0.2%
Bacillat® 35 preservative	0.1%

Ethylene glycol	64%
dye	4.0%
distilled water	31.2%
Potassium carbonate	0.5%
Tetrasodium ethylenediaminetetraacetate	0.2%
Bacillat® 35 preservative	0.1%

Diethylene glycol	45.0%
dye	3.0%
distilled water	51%
Potassium carbonate	0.4%
Bacillat® 35 preservative	0.1%

Triethylene glycol	45%
dye	3.0%

distilled water	51.4%
Potassium carbonate	0.5%
Bacillat® 35 preservative	0.1%

Ethylene glycol	62.0%
dye	4.0%
distilled water	33.4%
Potassium carbonate	0.5%
Bacillat® 35 preservative	0.1%

U.S. Patent Number 4,386,961
Heterologous Ink Jet Compositions
Lin
Exxon Research and Engineering Company
June 7, 1983

Propylene glycol	50–70%
Benzyl alcohol	8–50%
oleic acid	2–5%
Pontamine Black dye	20%
oleic acid	0–10%
Diethylene glycol	30–45%
Methoxy-triglycol	35–45%
Aldo MLD	0–7%
Lonza Pegosperse®	0–7%
dye	3–16%

U.S. Patent Number 4,421,559
Jet Printing Ink Composition
Owatari
Epson Corporation
December 20, 1983

dye	1.5%
glycerol	8%
Ethylene glycol	28%
urea	1%
distilled water	61.5%

dye	0.8%
gylcerol	13%
Ethylene glycol	20%
Potassium hydroxide	0.4%
urea	3%
distilled water	62.8%

dye	1%
Triethylene glycol	30%
urea	5%
distilled water	64%

dye	1%
Diethylene glycol	10%
Ethylene glycol	25%
Sodium hydroxide	0.5%
urea	1%
distilled water	62.5%

dye	0.8%
glycerol	10%
Ethylene glycol	20%
Potassium hydroxide	0.4%
urea	10%
distilled water	58.8%

U.S. Patent Number 4,986,850
Recording Liquid: Water Composition with 25 CP Minimum Viscosity at 9:1 Blend and 15 CP Maximum Viscosity at 1:1 Blend
Iwata et al.
Canon
January 22, 1991

dyes	4.8%
Diethylene glycol	15.0%
Polyethylene oxide-propylene oxide adduct of glycerin (viscosity modifier)	1.5%
water	78.7%

dye	3.0%
Diethylene glycol	20.0%
Polyethylene glycol	10.0%
1,3-Dimethyl-2-imidazolidinone	15.0%
water	52.0%

dye	3.0%
Diethylene glycol	30.0%
Polyethylene glycol 400	2.0%
water	65.0%

dyes	5%
Polyethylene glycol	35.0%
water	60.0%

II.2 PIGMENT BASED INKJET INKS

U.S. Patent Number 5,707,433
Pigment Inks for Ink Jet Printers
Kuge et al.
Fuji Pigment Co., Ltd.
January 13, 1998

Pigment — Carbon black	30.0 parts
Nikko Chemicals–Nikkol® DLP-10	10.0 parts
(Polyoxyethylene laurylether phosphate)	
Ethylene glycol	28.0 parts
deionized water	80.0 parts
ICI–Proxel® GXL preservative	2.0 parts

add after mixing to:

glycerol	50.0 parts
Diethylene glycol	150.0 parts
48% NaOH solution	2.0 parts
deionized water	648.0 parts

Pigment — Carbon black	30.0 parts
Dai-ichi Kogyo Seiyaku–Polysurf® A207H (Polyoxyethylene nonylphenyether phosphate)	10.0 parts
Ethylene glycol	30.0 parts
ammonia water	2.0 parts
deionized water	80.0 parts
ICI–Proxel® GXL preservative	2.0 parts

add after mixing to:

glycerol	200.0 parts
Propylene glycol	50.0 parts
deionized water	596.0 parts
Pigment–Red 122	30.0 parts
Nikko chemicals–Nikkol® DNPP-4	15.0 parts
(Dipolyoxythylene nonylphenyl ether phosphate) ethylene glycol	30.0 parts
ICI–Proxel® GXL preservative	2.0 parts
48% NaOH solution	1.0 parts
deionized water	72.0 parts

add after mixing to:

glycerol	200.0 parts
ammonia	2.0 parts
deionized water	648.0 parts
Pigment–Blue 15	20.0 parts

Kao–Emal® 20-T	15.0 parts
(Polyoxyethylene alkylether triethanolammonium sulphate)	
Ethylene glycol	20.0 parts
deionized water	98.0 parts

| ICI–Proxel® GXL preservative | 2.0 parts |

add after mixing to:

Diethylene glycol	200.0 parts
Ethyleneglycol dimethylether	10.0 parts
Triethanol amine	5.0 parts
deionized water	635.0 parts

U.S. Patent Number 5,316,575
Pigmented, Low Volatile Organic Compound, Ink Jet Composition and Method
Lent et al.
Videojet Systems
May 31, 1994

Pigment–Lithol Rubine Red	12%
Joncryl® 67 resin (20% in water)	50%
deionized water	35.8%
Surfynol® 104 (defoamer)	0.5%
N-methyl 2-pyrrolidine (solvent)	1.5%
Giv Gard DXN biocide	0.2%

Pigment–Alkali Blue 97013	8.0%
Joncryl® 57 resin	40%
water	39.8%
Giv-Gard DXM	0.2%
Surfynol® 104	0.5%
(1:1 in N-methyl-2-pyrrolidone)	
N-methyl-2-pyrrolidone	1.5%

Pigment–Carbon black	12.0%
Joncryl® 60	40.0%
water	45.8%
Giv-Gard DXN	0.2%
Surfynol® 104	0.5%
(1:1 in N-methyl-2-pyrrolidone)	

U.S. Patent 5,514,207
Low Molecular Weight Polyethylene Glycols as Latency Extenders in Pigmented
Ink Jet Inks
Fague
Xerox Corporation
May 7, 1996

| Pigment–Levanyl® A-SF carbon black | 20% |

sulfonate	20%
Polyethylene glycol 400	5%
Dowicil® 200 biocide	0.1%
deionized water	54.9%

Pigment dispersion–Hostafine® TS	4%
1,3-Dimethyl-2-imidazolidone	10%
Polyethylene glycol 600	5%
Dowicil® 200 biocide	0.1%
deionized water	80.9%

U.S. Patent Number 5,026,427
Process for Making Pigmented Inks
Mitchell and Trout
E.I. Dupont de Nemours and Company
June 25, 1991

Blue G XBT-5R3D pigment	500 grams
Daniels Disperse-Ayd W-22 dispersant	125 grams
distilled water	625 grams
Ethylene glycol	416.7 grams

Carbon Black pigment	12.5 grams
Triton X-100® dispersant	3.0 grams
distilled water	142.5 grams
Diethylene glycol	95.0 grams

Yellow YT-B56D pigment	150 grams
Tamol® SN dispersant	30 grams
distilled water	1056 grams
Ethylene glycol	264 grams

U.S. Patent Number 5,716,435
Recording Fluid for Ink-Jet Printing and Process for the Production Thereof
Aida et al.
Toyo Ink Manufacturing Company, Ltd.
February 10, 1998

Pigment dispersion	30 parts
Johnson Polymer Corp.–Joncryl® 62	3 parts
(acrylic resin aqueous solution)	
Kao Corp.–Emulgen® 420	2 parts
dispersing agent	
purified water	50 parts
glycerine	6 parts

Pigment dispersion	13.5 parts
Kao–Emulgen® 420 dispersing agent	0.2 parts
Gifu Shellac–Emapoly® TYN-40	3.0 parts
(acrylic resin aqueous solution)	

Ethylene glycol	30.0 parts
Sodium omadine antifungal agent	0.15 parts
Sodium ethylenediamine tetraacetic acid	0.02 parts
purified water	63.53 parts

Pigment dispersion	30.0 parts
Nippon Polymer Industry Co.–F-157	1.3 parts
(acrylic resin aqueous emulsion)	
Zeneca K.K.–Solsperse® 27000	0.5 parts
(dispersing agent)	
purified water	64.0 parts
glycerine	5.8 parts
Dimethylethanolamine	0.1 parts

Pigment dispersion	25.0 parts
glycerine	20.0 parts
Sodium omadine antifungal agent	0.015 parts
Sodium ethylene tetraacetic acid	0.02 parts
Diethylene glycol monobutyl ether	0.5 parts
Kao Corp.–Antifoam® E-20	0.8 parts

Pigment dispersion	13.5 parts
Kao Corp.–Emulgen® A-90	0.02 parts
(dispersing agent)	
Nippon Polymer Industry Co.-W-215	1.0 parts
(acrylic resin emulsion)	
Ethylene glycol	10.0 parts
Sodium omadine antifungal agent	0.15 parts
Sodium ethylenediamine tetraacetic acid	0.02 parts
purified water	63.53 parts

Pigment dispersion	13.5 parts
Nippon Polymer Industry Co.-W-215	1.0 parts
(acrylic resin emulsion)	
Diethylene glycol	20.0 parts
Zeneca K.K.–Proxel® GXL	0.15 parts
(antifungal agent)	
Sodium ethylenediamine tetraacetic acid	0.02 parts
Dimethylethanolamine	0.1 parts
purified water	63.53 parts

Pigment dispersion	55.0 parts
Johnson Polymer Corp.–Joncryl® 61J	7.0 parts
(acrylic resin aqueous solution)	
Kao Corp.–Emulgen® A-90	4.0 parts
(dispersing agent)	
purified water	56.0 parts
Diethylene glycol monobutyl ether	3.0 parts
N-Methyl-2-pyrrolidone	3.0 parts
2,4,7,9-Tetramethyl-5-decyne-4,7-diol	1.0 parts
Ethylene glycol	1.0 parts

Ejection Tests of Biological Fluids

The results from the following tests are not presented in order to document optimized recipes for jettable fluids carrying biological payloads, but as an illustration of the type of testing required to develop a workable ejection fluid. The descriptions of the ejection behavior of the various simple fluid mixtures are an illustration of the results one might expect from the ejection testing of generally nonoptimized fluid recipes. What we have observed is that developing fluids that are both chemically compatible for ones particular application and rheologically suitable for ejection a given droplet diameter from a particular type of drop ejector is not a trivial task. The following results may be considered to be both a warning and a reality check so far as the difficulty of optimizing biocompatible microdrop ejection fluids.

III.1 TEST SERIES I

Biological fluid ejection test results fluids supplied by professor Patrick O. Brown's Stanford University laboratory (January 27, 1999).

III.1.1 Test Fluids

- DNA marker 1 microgram/microliter, 12 kilobases
- DNA marker diluted to 10% concentration in distilled water
- Yeast cell suspension (shaken before loading)
- E. coli cell suspension
- Spe I enzyme in 20% glycerol

III.1.2 Test Procedure

Fluid ejection was tested from:

- 20 micron diameter micromachined nozzle ejector
- 50 micron ejection hole diameter glass capillary ejector

Parameters that were varied to attempt to obtain stable, satellite free ejection:

- Drop ejector internal pressurization
- Electrical pulse drive amplitude
- Electrical pulse drive pulse width
- Double pulse drive vs. single pulse drive
- Pulse rate

Restart behavior from a nonoperating state was also tested. Fluid volume in the drop ejectors were about 3 to 4 microliters. The fluids were loaded by being drawn in from the ejector tip under vacuum.

III.1.3 Results

III.1.3.1 DNA Marker 1 Microgram/Microliter

No drive setting was found with this fluid that would produce satellite free ejection of drops. The general ejection pattern, when tuned for best operation, was the production of an 18- to 22-micro diameter drop surrounded by a random spray of smaller satellites. Stroboscopic sequencing through the ejection process revealed that a long, thin fluid tongue (0.7 mm long) connected the primary drop to the dropper aperture as it was ejected. This fluid tongue would break up at about 0.7 mm length and fragment into a random spray of smaller droplets.

The amplitude needed to eject the DNA marker fluid increased as the ejection frequency was reduced. If the drop ejection process was stopped and the dropper allowed to sit for several minutes, the pulse amplitude required to restart drop ejection was about four times that required to eject fluid when being pulsed at 30 Hz.

This fluid was very hard to work with. It has a low surface tension that makes formation of the ejected fluid jet into a single drop an unstable nonreplicable process. The fluid acts as if it slowly evaporates and forms a viscous or solid boundary layer on surfaces exposed to the air. This makes drop-on-demand operation difficult to impossible if satellite free ejection is required. This is because a very high amplitude pulse is needed to clear the aperture after a period of nonejection. This large amplitude pulse unless of exactly the right drive level will throw out a large amount of spray in addition to the primary drop. Subsequent pulses are required to be of reduced amplitude appropriate for the frequency of operation.

Directional accuracy of the ejected jet, i.e., lack of angular skew, though was good.

III.1.3.2 DNA Marker Diluted (10% Marker, 90% Distilled Water)

This diluted mix ejected with about one third the drive amplitude that the full concentration fluid required. Satellite free ejection was attainable.

There was still a rate dependence upon the amplitude needed for stable droplet ejection. Excess amplitude will throw spray, so setting the correct drive amplitude is critical if this device is to be used for spotting an array. For some reason at below ejection rates 1 Hz stable satellite free operation was not possible. In one test, after the dropper was stopped for 30 minutes, restart required a pulse amplitude of ten times the 30 Hz steady state pulse amplitude.

The ejected drops had good directional accuracy.

III.1.3.3 Yeast Cell Suspension

Satellite free ejection was achievable. There was however a residual random jitter to the angular accuracy and drop ejection distance compared to tests done using pure water. This manifested itself in a worst-case drop ejection angle deviation of about 15 degrees from a perpendicular jet. There were also occasional clogs that would form which required that a drive pulse of about eight times the normal ejection amplitude be applied to clear the aperture and resume ejection. The time interval between random clogs was about 15 minutes.

This behavior might be explained if the yeast cells formed randomly sized clumps that would effect the ejection process by disturbing the fluid flow in the aperture, and if large enough in diameter, would block the aperture entirely requiring a high amplitude pulse to force the clump through or break it up.

The yeast cell tube was then left out over the weekend at room temperature by mistake. Fluid was drawn from this three-days-out tube and tested for ejectability. Nothing resembling stable ejection was achievable: either sprays were produced or the aperture would act as if it were clogged. The amplitude required for spray ejection from the three-days-out yeast suspension was about 20 times that which was previously required to eject the fresh yeast cell suspension.

III.1.3.4 E. Coli Cell Suspension

Satellite free ejection almost equivalent in stability to pure water was achieved. The directional and ejection velocity jitter observed was about a tenth that observed in yeast cells. No clogs were experienced. The pulse amplitude required for drop ejection was about one fourth that required to eject the yeast cell suspension. The main problem observed was that while the angle of ejection of the drops was stable, it was not necessarily perpendicular to the aperture face or repeatable upon draining and refilling. Differences in ejection angle were also observed upon changing the drive settings.

III.1.3.5 Spe I Enzyme, 20% Glycerol

This fluid formed the most stable satellite free drop streams of any of the test fluids including pure water. A very wide range of drive settings was able to produce directionally stable satellite free drops streams. No jitter was observed. Restart characteristics are excellent. The drive amplitude required is about twice that needed to eject pure water.

Glycerol and water as a carrier fluid appears to be a winner.

III.1.4 Summary of Results

We found out from inkjet technology literature and patent searches that commercial inkjet printer manufacturers had to develop rather complex ink formulations containing surfactants, detergents, and multicomponent carrier fluids designed to suppress aperture clogging, pigment agglomeration, and surface debris build up while having the required viscosity, surface wetting, and surface tension characteristics needed for reliable directionally accurate ejection. The surface coatings of the drop ejectors had to be matched to the fluids used.

A similar systematic series of tests may have to be conducted to determine whether we can find biocompatible carrier fluids for biological materials such that the resultant suspensions are accurately ejectable on demand from drop generators constructed from sterilizable biocompatible materials.

However, all the biological suspensions supplied to us in their standard solvent carrier fluids are currently ejectable if one is not concerned with directional accuracy of the drop stream or freedom from satellites and spray. This may be sufficient for the time being to test for alteration of the biomaterial by the ejection process.

The gross drop forming behavior of the different fluids were identical in both the glass aperture and the micromachined aperture drop ejectors. The micromachined apertures in general provided more reliable, lower drive amplitude ejection thresholds for drop generation than for the glass capillary ejectors. The differences in ejection threshold though were within a factor of two.

The all glass ejectors have the advantage of continuous visibility of the entire fluid column and the ability to observe the fluid to check for settling of suspended solids and the intake of bubbles near the aperture.

Directional stability of the drop stream was superior for the micromachined ejectors. This is because micromachined apertures have very smooth surface finishes and can have guaranteed axial symmetry. Glass apertures utilize mechanical grinding as part of their fabrication process. This mechanical grinding produces random chips and scratches along the aperture surface which act as symmetry breaking features that can misdirect a fluid jet.

III.2 TEST SERIES II

DNA suspension ejection tests. Test fluids supplied by Dr. Mary Tang, Stanford Nanofabrication Facility, Stanford University (March 24, 1999).

III.2.1 Test Fluids

- Control Buffer-10mM Tris-Acetate, pH 8.2
  - 10 mM (millimolar) indicates a concentration of 0.01 moles/liter
- Plasmid DNA in buffer
  - The concentration of DNA is 250 micrograms/milliliter
  - The aqueous solution is 10 mM Tris-Acetate

- MO reagent
  - The concentration of MO is 0.38 mM in 10 mM Tris-Acetate
  - MO reagent compactifies DNA
- DNA + MO
  - DNA concentration 250 micrograms/milliliter
  - MO concentration is 0.38 mM
  - MO packs DNA into particles that are 40 nm in diameter
- Polyplex (DNA + dendrimer)
  - DNA is at 200 micrograms/ml
  - Dendrimer is at 0.8 mg/ml
  - Dendrimer packages DNA into particles that are about 150 nm in diameter
- Dendrimer
  - Dendrimer alone at 1.6 mg/ml

The DNA is plasmid DNA. This is a double-stranded, supercoiled, circular loop, about 5k base pairs long (about 0.85 microns long, if you were to cut the loop, and lay it out straight.) A typical DNA for gene chips will be a linear, single stranded molecule, about 1000 bases long (about 0.34 microns long.) So, the DNA samples here represent the worst case in DNA solutions.

III.2.2 Test Procedure

- The drop ejector was a piezoelectrically driven glass pipette type dropper.
- The ejection aperture was a micromachined structure utilizing KOH etching to produce a pyramidal pit in a 0.5 mm silicon wafer terminating in a 20-micron square hole. The outside surface was silicon nitride. The aperture was attached by microwelding.
- An external manometer was used to control the interior pressure of the dropper in order to optimize ejection.
- Single pulse excitation was used. The default pulse width was 1.4 microseconds. If stable ejection was not achieved, then different pulse widths were tried. In practice however, if 1.4 microseconds did not produce stable ejection, none of the other settings would either. The drop ejection rates tested were from 1 Hz up to the limit where ejection instability set in.
- Where stable ejection could be achieved, stop and restart behavior was tested. By shutting down the drive pulses and then reapplying the drive at the same previous steady state amplitude, restart on demand ability was tested.
- The amount of fluid loaded was 3 microliters (2 mm fluid height).

III.2.2.1 Cleaning Protocol

- First the unejected remainder of the fluid in the drop ejector was forced out from the tip into a waste container by pressurizing the drop ejector.
- The cleaning consisted of four fluid fill and purge cycles using distilled water as the cleaning solvent.
- The distilled water was sucked into the tip under vacuum. This filling was done in an ultrasonic cleaning bath from a glass container holding distilled water. The amount of fluid drawn in was about five times that occupied by the previous test sample.

- This fluid would then be expelled in air from the tip under positive pressure into a waste container.
- After the last purge, the drop generator was attached to a vacuum pump to evaporate out the remaining water.
- This complete cleaning cycle took about 5 minutes to perform.

III.2.3 Results

III.2.3.1 Control Buffer-10mM Tris-Acetate, pH 8.2

Drive voltage	10.6 volts immediately after filling
	After running overnight, the drive threshold rose to 18 volts. The upper threshold for stable ejection 25 volts.
Pulse width	1.4 microseconds
Manometer	−14 cm higher manometer pressure produced unstable ejection
Drop diameter	17 microns

The fluid had extremely stable ejection characteristics and perfect stop-restart behavior. The low frequency operation of the dropper with this fluid was stable down to 1 Hz. The high frequency upper limit for stable drop ejection was approximately 500 Hz. At 300 Hz the ejected drop stream started to become directionally unstable.

III.2.3.2 Plasmid DNA in Buffer

Drive voltage	46.4 volts, stable up to 58.8 volts
Pulse width	1.4 microseconds
Manometer	Stable operation could be achieved over very wide ranges of pressure. The dropper could operate open to the air without a manometer attached.

Continuous, satellite free ejection was easily achieved. There was a slight amount of jitter to the drop positions when observed under synchronous strobe illumination. Jitter along the direction of ejection was about 0.2 millimeters. Lateral jitter small and difficult to determine.

The maximum stable drop ejection rate was 1.4 kHz. Ejection was stable down to 1 Hz though the drive voltage increased slightly below 2 Hz.

Instant restart was possible after a 70-minute period of shut down operation. Longer intervals were not tested.

III.2.3.3 MO Reagent

Drive voltage	15 to 20 volts, had to be constantly varied
Pulse width	0.7 to 5 microseconds, no stable operating point
Manometer	Varied from 0 cm to 40 cm, no stable operating point. Some negative pressure was mandatory to produce satellite free ejection.
Drop diameter	16 to 18 microns

Satellite free ejection could be achieved but not for more than 15 seconds at a time until the drive amplitude had to be readjusted. The directional stability of the ejected drops was poor.

At ejection rates greater than 300 Hz there was a rapid fluid build up on the surface of the aperture over the ejection hole. Stopping the drive pulses and applying high negative pressure resulted in the fluid being drawn back into the dropper.

When expelling the fluid from the tip of the dropper after the test was concluded, a 1 cm diameter bubble was formed as air was blown into the meniscus remaining in the dropper tip. There was evidentially a very large reduction in surface tension caused by the MO reagent.

III.2.3.4 DNA + MO

Drive voltage	22 volts, varied to attempt to stabilize the drops
Pulse width	0.7 to 5 microseconds, no stable operating point
Manometer	No stable operating pressure
Drop diameter	14 to 18 microns, uncertain due to the difficulty of sustaining a continuous drop chain

Satellite free ejection could only be achieved for ten second intervals before retuning was needed. Ejection stability was very poor.

Restarting the dropper could not be done reliably. Partial drive amplitude pulsing (reducing the drive amplitude to a point below ejection threshold, but enough to cycle fluid in and out of the ejection hole) did not help restart reliability.

III.2.3.5 Polyplex (DNA + dendrimer)

Drive voltage	11.4 volts
Pulse width	0.7 to 6 microseconds, no stable operating point
Manometer	−15 cm to −40 cm
Drop diameter	14 to 18 microns, uncertain due to difficulty sustaining a continuous drop chain

There were no stable settings. A lot of satellites accompanied the primary drops. Drop ejection patterns changed over periods of 30 seconds. The ejected drops had extremely poor directional stability. Low frequency operation was extremely directionally unstable (ejection angle variations of up to 30 degrees.)

III.2.3.6 Dendrimer

Drive voltage	12 volts
Pulse width	1.4 microseconds
Manometer	Functional from −10 to −20 cm
Drop diameter	22 microns

Satellite free operation was achievable at slow ejection rates (<30 Hz.) Past 30 Hz the ejection went unstable due to surface fluid build up on the face of the ejection aperture.

Directional ejection varied with amplitude but was consistent for a given setting.
The fluid ejector combination was capable of having drop ejection stop and then
restarting on demand.

III.2.4 Supplemental Control Tests Utilizing the Same Drop Ejector

III.2.4.1 Distilled Water

Drive voltage	18.6 volts
Pulse width	1.4 microseconds
Manometer	−3.5 cm
Drop diameter	16 microns

Stable satellite free drops were immediately produced. The drops were stable
from 1 Hz to 1 kHz. Drop production could be stopped and restarted on demand.

III.2.4.2 Distilled Water with 20% Glycerol

Drive voltage	29.6 volts
Pulse width	1.4 microseconds
Manometer	+18 cm for best start-stop behavior. Reliably produced drops at wide ranges of negative pressures.
Drop diameter	24 microns

The drop streams had good stability with no satellites. Rate 1 Hz to 2.1 kHz.
The drop streams could be stopped and restarted on demand.

III.2.4.3 Distilled Water with 20% Propylene Glycol

Drive voltage	25.2 volts
Pulse width	1.4 microseconds
Manometer	−20 cm
Drop diameter	22 microns

Stable satellite-free ejection was achieved from 1 Hz to 900 Hz. Instant restarts
were achievable with a slight first drop jitter in the restarted drop stream.

III.2.4.4 Fluid #2 (Plasmid DNA in Buffer) with 20% Glycerol

Drive voltage	60 volt range
Pulse width	0.7 to 5 microseconds, no stable settings
Manometer	no stable settings
Drop diameter	15 to 25 microns (estimate based on optical image)

The drops produced had very poor ejection consistency and very poor directional
ejection stability.

One big problem observed was continuous fluid build up on the outside of the drop ejection aperture over the ejection hole while ejecting drops. Negative pressure up to –60 cm does not suppress this. This surface fluid build up appeared to be the major source of the drop ejection instabilities.

III.2.4.5 Fluid #2 (Plasmid DNA in Buffer) with 20% Propylene Glycol

Drive voltage   50 to 60 volts
Pulse width   0.7 to 5 microseconds, no stable operating point
Manometer   0 to –60 cm, no stable operating point
Drop diameter   15 to 25 microns (estimate based on optical image)

The drops had very poor ejection characteristics along with very poor ejection directional stability.

Fluid build up over the ejection aperture hole was a problem. Starting and stopping the dropper required physically wiping the aperture to get the drop stream to restart.

III.2.5 Summary of Results

The best ejection fluid for DNA was fluid #2, the Plasmid DNA in 10mM Tris-Acetate buffer. It had excellent drop stream stability and restart capability. This is in contrast with the DNA marker fluid initially provided by professor Brown's lab. This fluid, described as a 1 microgram per microliter concentration, 12 kilobase pair DNA marker, was incapable of being ejected without satellites and which could not be reliably restarted once continuous ejection was stopped.

The Tris-Acetate buffer solution ejected better than pure distilled water. Adding the DNA to the Tris-Acetate buffer increased the pulse voltage level needed for ejection and introduced a slight jitter to the drop-to-drop ejection positional accuracy. Most (possibly all) of the jitter was in the direction of the drop travel and not ejection angle deviations. The stop-restart behavior was excellent.

The DNA compacting additives, the MO reagent and the dendrimer made drop ejection unreliable and restart on demand impossible. Tests performed ejecting these DNA compacting agents alone, indicated that this unstable ejection behavior is probably due to the unfavorable rheological characteristics of these additives. In particular, the MO reagent drastically reduced the surface tension and contact angle of the solution to the point where surface wetting of the aperture resulted in a fluid build up large enough to destabilize drop ejection. What would be desirable would be to try using DNA compacting agents that do not alter the rheological characteristics of the solution.

Adding propylene glycol and glycerol to the Plasmid DNA in buffer solution made ejection characteristics much worse. For some reason adding the humectant caused fluid to build on the outside of the ejection aperture hole. This fluid build up was severe enough as to require physically wiping the outside of the aperture place in order to restart the dropper if the electrical drive pulses were stopped.

There is a slight mystery here. A 20% mixture of glycerol in water ejects quite well. The DNA in buffer solution ejects well. However adding glycerol to the DNA in buffer solution (enough to make a 20% mixture) produced a fluid that could not be made to stability eject. Why would additives to water that when added individually be ejectable, but when mixed, produce an unejectable solution? In addition this combined fluid mixture seemed to be failing to reliably eject due to it having an extreme tendency to wet the outside of the aperture. This is a behavior that was not observed with either of the individual fluid mixtures.

Index